Ethics
in Science

Ethical Misconduct in Scientific Research

Ethics in Science

Ethical Misconduct in Scientific Research

John D'Angelo

CRC Press
Taylor & Francis Group
Boca Raton   London   New York

CRC Press is an imprint of the
Taylor & Francis Group, an **informa** business

CRC Press
Taylor & Francis Group
6000 Broken Sound Parkway NW, Suite 300
Boca Raton, FL 33487-2742

Printed in the United States of America on acid-free paper
Version Date: 20120229

International Standard Book Number: 978-1-4398-4086-3 (Paperback)

Library of Congress Cataloging-in-Publication Data

D'Angelo, John.
    Ethics in science : ethical misconduct in scientific research / John D'Angelo.
      p. cm.
    Includes bibliographical references and index.
    ISBN 978-1-4398-4086-3 (pbk.)
    1. Research--Moral and ethical aspects--Case studies. 2. Fraud in science--Case studies. I. Title.

Q180.55.M67D36 2012
174--dc23
                                            2012004944

Visit the Taylor & Francis Web site at
http://www.taylorandfrancis.com

and the CRC Press Web site at
http://www.crcpress.com

This book is dedicated to all of those who have supported me in everything I've done. Without each of you, I am nothing.

Contents

Introduction

Just about every year, we hear about another case of ethical violations in some field of science. None of the many fields of science are immune from this betrayal. The reasons for these violations of scientific ethics are manifold, and a discussion of these reasons, along with discussions of what constitutes a violation of scientific ethics and how the scientific community detects them, is provided in Chapter 1. Certainly, one can argue that these issues cannot be resolved without instituting a formal moral code and that, because nearly every one of us has our own individual moral code, such initiations cannot be made. However, for science to survive, we must agree on a basic set of appropriate behavior, at least in principle. Indeed, an overwhelming majority of active and retired scientists have agreed upon and follow such a basic set of appropriate behaviors. Unfortunately, as with any portion of society, there are a few dissenters. These dissenters vary from those who feel the rules do not apply to them, to those who disagree with the rules, and even all the way to those who actively and consciously try to "beat the system." Perhaps most troubling, however (only because this group's actions could easily have been prevented), are the instances where the perpetrator honestly *did not understand* that his or her actions were a violation of scientific conduct. Other troubling cases (that are more difficult to address) are where the system actually facilitates, rather than inhibits, the violations.

Although this work considers the potential misconduct toward human or animal research subjects, it should be noted that it is not the intended spirit of this work to delve into a discussion of the "right" and "wrong" of issues such as human and animal testing, cloning, and stem cell research—though they will receive brief mention. These are deeply personal and, in many instances, religious issues, and I am not one to tell you how to feel about them. A discussion of these current issues, especially with respect to the government's role in providing funding for such research and popular opposition, is in Chapter 5. Instead, it is the intended spirit of this book to focus on improper conduct within scientific research.

In Chapter 8, various cases involving potential misconduct in science are presented. Every attempt has been made to adequately represent both sides of the issue when possible. This has not been done as an attempt to defend or justify the actions of the putative violator, but instead as an attempt to provide maximum insight into the issue so that something can truly be learned from each case. In each of these cases the year and location of the offense (or alleged offense) is provided, along with the names and claims of all of the principal players. If a resolution has been reached in the case, it is also provided. This format allows for several advantages. First, it allows the reader to observe how widespread, both in field and in geographic location, misconduct in science has been, demonstrating that nobody is immune. Ethical violations occur in every country, are committed by people of every race, and are committed (at least allegedly) by even the most famous of researchers. Second, it allows the reader to see all the sides and factors into the issue. This has the effect of giving a real firsthand look at why such ethical violations occur. Finally, when possible, the resolution allows the reader to see what happens to the offenders, to those negatively affected by the offense, and the whistle-blowers (the persons who call attention to the offense). Also, some of the cases are presented specifically because they are tremendously muddied and there has not yet been a resolution—and with none on the horizon, either. It is the desired outcome that these latter cases, especially, stimulate discussion in the classroom.

If the cases of scientific misconduct where true ignorance is the cause can be reduced, progress will have been made. Such education is one of my goals here. It is also a major goal of the American Chemical Society, which has incorporated a section on science ethics into the third edition of the *ACS Style Guide*. Hopefully, the discussions that result from this book will enlighten students to what sorts of behaviors constitute scientific misconduct. An additional goal of my book is to show that science is in large part self-correcting and self-policing, and that it is truly not worth trying to get away with any sort of scientific misconduct.

How to use this book

It is suggested that this book be used the following way

After reading Chapters 1, 2, and 7, the case study can easily begin. I suggest that students write no more than a one-page summary of their thoughts on each case, whenever the case study begins. Also, the instructor should not feel bound to only use cases herein. After an instructor-mediated group discussion in class, another one-page summary by the students about how their minds were either changed or supported by the discussion should be completed. This may allow for a course to be built around this concept to help satisfy a public speaking requirement at some universities. In instances where the instructor would like to build a writing requirement into the course, there is room for flexibility to increase the demands on student writing.

An important note on Chapters 3 through 6

Chapter 3 does not formally cover scientific misconduct. It does, however, address factors that I feel can contribute to scientific misconduct, at least indirectly. Chapters 4 through 6 cover increasingly important issues in science today. It is for this reason that I have included them in this book, detached from the discussion about misconduct in science. They are provided for those readers interested in their respective topics, though Chapter 8 of this book can certainly be used without having read Chapters 4–7.

About the author

John D'Angelo, PhD earned his Bachelor of Science degree from the State University of New York at Stony Brook in 2000. He went on to earn his PhD in chemistry from the University of Connecticut, where he worked in the laboratories of Professor Michael B. Smith in 2005. After a 2-year postdoctoral research assistantship in the laboratory of Professor Gary H. Posner, Professor D'Angelo secured his first tenure-track assistant professor position at Alfred University, Alfred, New York, where he remains today. His research interests involved various pedagogical strategies including ethics in science and introducing more exciting lab experiments into the organic chemistry lab curriculum, in particular labs that are interdisciplinary. Laboratory-based research interests include the synthesis of small-molecule fatty acid synthase II inhibitors and further developing the use of the conducting polymer Poly ([3,4-ethylenedioxy]-thiophene) as a chemical reagent.

chapter one

Irresponsible conduct in research

What is it, why does it happen, and how do we identify it when it happens?

We should start by finding out what the dictionary has to say about the word *ethics*. And so, according to www.dictionary.com, ethics (as it applies here) is defined in the following way:

1. A system of moral principles.
2. The rules of conduct recognized in respect to a particular class of human actions or a particular group, culture, etc.

Armed with a definition of ethics, we can then ask: What does this mean to science? Unfortunately, the answer to this question is often very unclear. Although scientific ethics is almost certainly more #2 than #1, definition #1 almost always asserts itself in individuals. Despite how open to interpretation many ethical issues are, the scientific community has by and large agreed upon a standard of behavioral principles to which the vast majority of practicing scientists adhere. As with every walk of life, however, the exceptions receive most of the attention and ruin it for those of us who are honest. The baffling paradox is that, more often than not, the offenders get caught red-handed and red-faced, ruining their careers. If this is the case, why do they do it? This very question is explored later in the context of each violation of scientific conduct as the chapter develops.

Not surprisingly, the professional organizations in all of the science fields have established guidelines for their members' ethical behavior. Some of these guidelines can be found online at the following websites:

- American Chemical Society's Chemist's Code of Conduct[1]
  - http://www.acs.org, follow the "careers" tab in the ribbon, then follow the ethics and professional guidelines link (accessed June 2011).

---

[1] This is also called the Chemical Professional's Code of Conduct.

- American Institute of Chemical Engineers' Code
  - http://www.aiche.org/About/Code.aspx (accessed June 2011).
- Society for Integrative and Comparative Biology
  - http://www.sicb.org/about/code.php3 (accessed June 2011).
- National Society of Professional Engineers
  - http://www.nspe.org/Ethics/index.html (accessed June 2011).

This is by no means an exhaustive list. Nearly all professional organizations make such guidelines available somewhere, at least to their members.

Another issue that must not be ignored and is covered at the end of this chapter is what happens to the person who calls out the offending scientist? However unfortunate the moniker may be, this person is usually referred to as a whistle-blower. Also included is the highly important distinction between the triple forces of bad ethics, bad science, and genuine errors that must be made.

What constitutes scientific misconduct?

Before embarking on a discussion of why ethical violations in science occur and before ultimately performing a case study of ethical violations in science, we must first identify just what scientific misconduct is. Motivations are different for each "crime," and you simply cannot determine why people do something wrong or how to prevent it if you do not know what they are doing wrong. Ethical violations can be committed many ways. Several specific violations are presented below.[2] They are:

- Intentional negligence in the acknowledgment of previous work (including work *you* did)
- Deliberate fabrication of data you have collected
- Deliberate omission of known data that does not agree with the hypothesis
- Passing another researcher's data as one's own
- Publication of results without the consent of all of the researchers
- Failure to acknowledge all of the researchers who performed the work
- Conflict of interest
- Repeated publication of too-similar results or reviews
- Breach of confidentiality
- Misrepresenting others' previous work

[2] The specifics of this sort of list may vary from source to source. The content is ultimately the same, regardless. This list has been arranged in this way in an attempt to make the discussion more fluid.

What these violations are, why they happen, and how they are caught will now be discussed. When appropriate, the damages each causes will likewise be covered.

Intentional negligence in acknowledgment of previous work

What is it?

First and foremost, this is not to be confused with the unintentional negligence in the acknowledgment of previous work done, which can be called poor science (something that will be touched upon later). In this instance, a researcher is intentionally not mentioning work that has already been done in the field. This does not necessarily mean *no* previous work is mentioned. The same violation is occurring when an author only acknowledges work inferior to his or her own. This is unethical because it then appears as if the author is the pioneer or leader in the field. When this occurs in a grant application, it may result in misappropriation of funds by the funding agency. However, in a grant, it may also cause the work to be viewed as having no precedent and therefore not a safe investment for the funding agency. This can therefore be a double-edged sword. Failing to acknowledge previous work also provides no context for the reader; that is, he or she cannot compare the present work to what has happened in the field in the past.

It can certainly be argued that it is not possible to track down every bit of work that has been done on a topic so that you can properly acknowledge all of the work that was done prior to your contribution to the field. With the enormous volume of work being done today,[3] this is certainly a daunting—if not impossible—task. Fortunately, however, there are electronic search engines that greatly facilitate these searches. Although the databases that these search engines access are certainly neither complete nor infallible,[4] an honest researcher will utilize these tools to the best of his ability. It is the researcher's absolute responsibility to make at the very least a good faith effort to search all of the relevant literature before proceeding with a publication or grant. Does this mean that you have to read every published work in every language? Certainly not! Nor does it mean a physicist must read the table of contents of every biology journal to ensure that he or she is working ethically. However, if the very subject you are attempting to publish or study is the title of an article that is in a language you are unable to read, you had better find a translator if you want to proceed ethically. Frankly, you had better find a translator in

---

[3] There are literally hundreds of journals for each of the sciences, many written in more than five different languages.

[4] If a typo is made in entering the paper or book into a database, you'll never find it even if you search for it properly.

order to ensure that you are not trying to reinvent the wheel, something that is much more like bad science than bad ethics.[5]

Why does it happen?

You may be wondering what someone has to gain by doing this, so a small divergence is necessary here to provide some background. There is great prestige with having your work cited many times. This implies that your work is important or at least well read. As a highlight of this importance, a relative measure of this has even been developed called the H-index. This H-index is a number that counts the number of references that have been cited that number of times. It is calculated in the following way: If a researcher has six publications that are cited six times each, he has an H-index of 6. If on the other hand, the researcher has ten publications that are each cited five times, he has an H-index of 5. If original research is not properly acknowledged, the wrong people get credit for pioneering a field, and this can have disastrous repercussions, including eventual misappropriation of Nobel Prizes. In short, the stakes are extremely high to climb the mountain first. This violation essentially gives a false impression of superiority. Even though the incentives are big to make it look like you were the first to climb that mountain, that is often a fruitless effort as it is by far and away the easiest ethical violation to catch. If someone leaves out work that was previously done, especially if it is in a field that is very competitive, it gives the appearance that the offender's method may be the best or only one, preventing the work from being viewed in any context. This is, of course, downright misleading and in a way gives false credit. This putative false credit then has the effect of making offenders look better than they really are and therefore potentially more likely to get a grant funded, to have the manuscript accepted for publication, or to have the research cited more times. Furthermore, there is a difference between not mentioning *any* competitors and not mentioning only the best competitors. It is absolutely a violation to only mention competitors you are superior to and to leave out those who are superior to you. The only proper way to proceed is to show where your work fits into the big picture.

How is it caught?

Ordinarily, this issue is caught by the peer-review process and is an instance where peer review works well, as far as detecting scientific misconduct goes. The peer reviewers of grants and other publications are often, if not always, experts in the field, at least in principle and almost always in practice. They are intimately familiar with the state of the art and who has made which contributions to the science. Consequently, if a reviewer knows of other work that ought to be referenced or included into

---

[5] The difference between bad science and bad ethics is discussed later in this chapter.

the discussion, she is obligated to indicate so in the review and usually will indeed do so. If the editor takes the review to heart, he will inform the authors that the appropriate references must be added, unless the authors can defend why those specific references should not be included.

Deliberate fabrication of data you have collected

What is it?

The more colloquial way to describe fabrication of data is lying. It is one instance of scientific misconduct that is also, without question, a violation of the moral ethics of decent people. In this ethical violation, researchers quite literally make up data, claiming experiments were carried out when they were not or significantly altering the results they do obtain so that they fit the pre-experiment hypothesis or previous studies. This ethical violation is among the most difficult to catch, if not the most difficult, because data fabrication can only be caught if another scientist attempts to repeat or otherwise use the research that was fabricated. This ethical violation is also perhaps the most damaging of them all—not just because it is undeniably, morally wrong, but because it also has the unfortunate consequence of leading other researchers down an incorrect and potentially impossible path trying to repeat and/or use the fabricated results. This then has many negative effects on other researchers' careers, resulting in wasted time and research funds. In these days of intense pressure to publish and to obtain grants in times of tight financial stress, losing time or money could result in grants being terminated or denied funding, or in the destruction of a new faculty member's chances of being awarded tenure or promotion. Consequently, fabrication of data violations is often met with severe repercussions, such as termination from a position or a ban on applying for federal grants. Even with such serious threats hanging overhead, many people still succumb to the temptation.

Also falling into this category is alteration of results (especially spectroscopic results such as nuclear magnetic resonance, NMR) to make the data appear more like that which was expected or desired, or to make the product of a chemical reaction appear more pure. Although some may argue, for example, that "It's no big deal, I only Photoshopped out the solvent," it is still scientific misconduct. It is most unfortunate that this is virtually impossible to catch. Many would argue that this is not as bad as quoting a crude yield of 96 percent while neglecting to report that the pure yield is 34 percent, and perhaps that is true ... though I cannot agree. These are, without question, bona fide fabrications of data. With the advent of more and more advanced computer programs, this type of data fabrication is becoming easier to accomplish with each passing year. In fact, the *Journal of Biological Chemistry* recently announced its adoption of the *Journal of Cell Biology*'s policy because this has become a more prevalent issue that reviewers and readers must be more cognizant of than ever before.[6]

Fabrication of data is significantly different from data that is accused of being erroneous but not fabricated. I would argue that these sorts of cases are neither an example of bad science nor bad ethics but instead an example of scientific progress. As a brief example, consider the ancient belief that the Earth was at the center of the universe. Based upon the evidence that was collected, this was indeed a logical conclusion for ancient man to make. Modern evidence refutes this, however, and we now know this to be incorrect. This does not make ancient man's approach toward this conclusion unethical. It also is not bad science. They took the data they had and made what they believed to be the most logical conclusion. As science has progressed, we have become capable of proving the Earth is not only *not* at the center of the universe but also not even at the center of the galaxy or even our stellar neighborhood, the solar system. This is clearly a case of scientific progress and our ability to both collect and interpret data. Examples like this abound and are neither bad science nor scientific misconduct.

Why does it happen?

Why do people fabricate their data? Well, this one should be really easy, even for someone who has nothing to do with any scientific field to answer. Many reasons should come to mind instantaneously. First, it is exceedingly difficult to catch, making it (potentially) very easy to get away with. Second, you will very rarely, if ever, get a grant or publish a manuscript based on poor results—and without grants and publications, you're not going to keep your job for very long and certainly have a difficult time earning tenure, a promotion, or a raise. Excuses for a pharmaceutical company to fabricate data are even clearer: Many millions of dollars hang in the balance. The incentive to display good results is clear; your career and livelihood depend on it. The greed to be (or appear to be, more accurately) the best is a very tempting motivation for some people.

How is it caught?

This is one of the worst forms of ethical misconduct and one of the hardest to catch. It is simply impossible to catch this via the peer-review process, as that would require a reviewer to check every claim.[7] With the enormous volume of work and increasing specialization of research, such a

[6] *Journal of Cell Biology*, 2004, 166, 11–15.
[7] Some journals, such as *Organic Synthesis*, do exactly this and are among the highest-profile journals as a result. However, detecting fabricated results is not the goal of *Organic Synthesis*.

widespread effort is nothing short of unreasonable. Wholesale fabrication of data is impossible to catch before publication. Usually, instead, this violation is learned the hard way by an innocent researcher trying to use or further develop the fabricated results.

One way to potentially catch the digital alteration of results may be to take advantage of the improving computer technology that the perpetrators exploit and have raw data files (that have the appropriate time and date stamps) sent to reviewers. The reviewer could then use the appropriate software to recreate the figures and perform a check. This, however, would put enormous strain on the peer-review process and, frankly, the point of peer review is to evaluate the science, not detect fraud. The science is taken as being true, and this trust is essential to science.

Deliberate omission of known data that does not agree with hypotheses

What is it?

This is very similar, but not identical, to the fabrication of data, especially since it may be better thought of as selective inclusion of data. If any particular results are left out of a publication just because they do not agree with pre-experiment hypotheses, an ethical violation has certainly been committed. We are obliged as scientists to report *all* the data (without revision) that we obtain. (However, rounding to the appropriate number of significant digits is absolutely acceptable.) Unfortunately, all too often there is a tendency to omit outlying data. Of the violations considered in this book, this may be the one for which coherent arguments can be made to defend it. That being said, this becomes a more egregious breach of code when statements in the publication suggest that a failed or omitted experiment did, would, or even should work. Such a hypothetical scenario is discussed later in this chapter.

There are tests, such as Student's t-test, the Q test, and confidence limits, among others, that allow a researcher to determine whether a data point is insignificant. In these cases where a data point fails any of the above-mentioned tests, it is entirely appropriate to omit the offending data. It would be appropriate, however, to mention the data that have been omitted from the conclusion and the grounds on which they were omitted. Although such exclusions may indeed be experimental or equipment error, they also may represent exceptions to your conclusion that may lead to a new direction of research. In cases where one data point from an experiment that was repeated ten times is eliminated, these tests are particularly applicable. However, they are *not* applicable to disqualifying a single drug candidate when nine others are active and the one outlier is significantly less active.

Why does it happen?

This is perhaps the only ethical violation that is defensible. It is most unfortunate that the current peer-review process for the publication of manuscripts and funding for grants is sometimes not really kind to aberrant data. Even in cases where the aberration is explainable, it is almost less problematic to leave it out altogether because often the reviewers red-flag the data and use them as grounds to reject the paper or grant because, for example, the process is not general enough or the new class of inhibitors has too many derivatives that are not good enough. Unfortunately, this very practice of reviewers acts not only to facilitate dishonesty but also to hold science back. Therefore, many times, you have to be dishonest (although perhaps "not fully disclosing" is more accurate) in this way to succeed. Widespread application of these sorts of reviews enables a culture where only the best results are published, rather than all of them. Full honesty is still the best policy, however. The more diligent reviewers, though not all, require these results to be included and will not hold a negative result against the author. Full disclosure only shows the limits of a method, a demonstration that is easily as important as the method itself. In fact, it is far better for you to show the limits of your own research than for someone else to show the limits. At least in the former case, you can provide the reasons for the "shortcomings" of your own work.

How is it caught?

This is almost impossible to catch; if researchers don't report that they did something, there is no way to know that they did. The only way that this can be caught is if a diligent reviewer thinks of the very experiment or example that has been left out and demands it as additional work that must be done before the publication can proceed. This sort of demand does occur. You may be wondering how this is not in conflict with the earlier statement that science is based on trust. Such a review is indeed an evaluation of the science; a reviewer can suggest additional experiments that would further confirm or refute the authors' claim. Thus, in such cases, it is done to help improve the science, not check it.

Passing another researcher's data as one's own

What is it?

A gross ethical violation is the passing off of someone else's work as your own. This is similar to, but different enough from, what was encountered in an earlier ethical violation (intentional negligence in acknowledgment of previous work) to warrant its own class. This falls into the category of plagiarism. One variety is where a researcher reads a paper or grant, or attends a seminar, and then attempts to publish the data *as is,* as his own work, while doing nothing more than changing the names and contact information of the authors. This is clear "highway robbery" and when encountered must be dealt with quickly to allow the credit to be placed in the right place. This specific mechanism is exceedingly rare.

This does not have to be limited to publications that researchers might review or seminars they attend. Ideas can be stolen as well, and to do so is as bad as stealing the accomplished work. Very rarely does an author acknowledge in any way a personal communication that led to a break-through, though authors always should.[8] The offender could also just as easily (and just as unfortunately) take ideas from a grant she refereed and incorporate those ideas into her own grant or research for her own gain. In other cases, the work being stolen is hastily repeated and then the new results written up, published, or incorporated into a grant. In the latter case, the fact that the work was repeated really does not make it accept-able. In fact, it perhaps makes it worse as it demonstrates additional effort toward being dishonest. A prime example of this was brought to light in 2006 in the case of Armando Córdova of Stockholm University in Sweden, discussed in Chapter 8 of this book. Also, with the enormous number of scholarly journals in print today, it is even possible to just resubmit someone else's work *as is* to another journal for publication with different names on it even after the original paper has been published. Millions of articles are published every year and it is truly impossible to check that many, despite the leaps in computer-assisted searching.

The question that you should be tempted to ask at this point is: Just what needs to be referenced? This is where "common knowledge" comes into play. For example, nowadays, we do not need to reference or cite in any way that all known life is carbon based. This is a fact that once was novel research and had to be discovered, is now common knowledge, and no lon-ger needs a reference. Similarly, nobody in his right mind would cite that high cholesterol causes heart disease, even though it was original research that revealed that fact long ago. Likewise, nobody references Watson and Crick's manuscript when describing the double-helical structure of DNA anymore. A good rule of thumb is that if the general public (nonscientists) knows it, it can safely be considered common knowledge and does not need a reference. A better rule of thumb is: When in doubt, cite!

Regarding presentations, seminars contain varying amounts of as-yet unpublished results. Occasionally, someone in the audience may think that something that is not yet published is publishable as is. She then may feel that if she could repeat the results, quickly, in her own lab, she might be able to scoop the original researcher and appear to get there first. Perhaps she might just publish the results directly from the seminar. This

---

[8] For example, an idea that comes from a discussion in your office with a colleague or visit-ing seminar speaker is considered personal communication.

is most unfair, as seminars are often given in confidence that the audience will not perpetrate such acts, and it is a horrendous breach of professional trust. Furthermore, department seminars are usually considered to be personal or private communications. Seminars that are at national meetings are an altogether different conversation. These are certainly a public disclosure, and they must be referenced in the same manner a journal article should be. It is therefore completely inappropriate to attempt to report data from someone's seminar as data you have collected, even if you repeated the work.

You may be wondering what happens in the rare cases that researchers make a simultaneous discovery. This can lead to a number of different scenarios, some of which are ethical, the others unethical. In the ethical case first, we assume that the reviewing researcher is also pursuing the same goal and is preparing a manuscript of his own. This is a reasonable assumption since most peer reviewers are experts in a particular field. In some fields, there are very few experts, so it is inevitable that a reviewer will also be submitting a publication reporting the same (or perhaps in some cases, different) conclusions. From here, there are a number of different ways the reviewer can react. First, and probably most ethically, the reviewer would alert the editor to this conflict and recuse himself, explaining the source of the conflict. The editor may then ultimately propose the papers be published in series with one another while being reviewed by someone else, since they are so closely related. In this way, neither researcher receives any significant benefit or detriment. However, it is important that the reviewer informs the editor *before* actually reading the paper because the reviewer cannot "unread" the paper and may inadvertently use this data in his own research, which *would be* unethical. If the reviewer only realizes the close relation after reading it, he must inform the editor at once.

Undoubtedly it is inevitable that a researcher will receive inspiration for a new project or critical new insights for an ongoing project during the review of a manuscript or grant submission. This is different from what was previously discussed, and these differences deserve mention. Here, it is the insinuation that the reviewer recognizes something not directly related that can still be applied to his own project. The case of reviewing a manuscript involves much more clearly defined lines and will therefore be considered first. The most appropriate way to proceed, if a reviewer of a manuscript identifies ways in which the work can be applied to his own work, is for the reviewer to immediately make this potential conflict of interest known to the editor. However the editor chooses to proceed (whether the reviewer is replaced or allowed to continue), the reviewer simply cannot unread what he just read and learned. This immediately opens up the potential conflict of "How long must I wait to act on what I read?" In all likelihood, the only ethical way to proceed is to wait until the manuscript has been published or otherwise been made available to the general public.[9] The *most* ethical way to proceed would be to even have the self-discipline to not do any work using the results in the manuscript until then. This removes any unfair advantage the reviewer would have potentially gained. However, many would likely consider that level of rigor overkill. Another potentially viable option may be to ask the editor for permission to contact the author and perhaps begin a collaboration based upon this potential common interest. It is critical to note that this sort of contact must have the editor's approval, if it can be done at all, since reviews are ordinarily anonymous. After it is published, it is fair game to contact the author. Even in an open review format, the editor should be consulted before a reviewer reaches out to an author in this way.

The situation is far more complicated when the reviewed work is a grant. Unlike nearly all manuscripts, which by and large are complete projects or at the very least complete mini-projects, grants are *proposed* projects, even though they often build upon previous work. This not-so-subtle distinction makes it much easier to steal work from a grant since journal articles (manuscripts) usually serve to "stake claim" to a field, something grants simply do not do in any way. Furthermore, publications can be searched for, while unfunded grants can be less easily found and no official record of the proposed idea survives. Because of this, and because grant reviews are absolutely anonymous, there is no ethical way to use novel information found in a grant. Even contacting the author is absolutely inappropriate. The only ethical way to proceed is to be patient. The situation can become even more complicated if, while reviewing the grant proposal, a reviewer envisions an equally novel but superior quality idea based upon the same work. This is likely to be inevitable and, since so much work goes into authoring a grant proposal, if the reviewer then submitted his superior idea in the next round of proposals, it would be very difficult to convict him of scientific misconduct. However, sometimes the court of public opinion is even more draconian than the real penalties, and this sort of behavior will certainly be met with anger from persons in the scientific community.

A related concern that applies to both manuscripts and grants is that the reviewer may come across a reference he never knew about. As before, this cannot be unread. Reviewing grants and manuscripts is more than just a service to one's field; it is also a way to learn. There would indeed be nothing unethical at all with using this newfound reference in the reviewer's own work.

In another related scenario, the reviewer may identify a way in which he can either improve the research under review or use the research to improve one or more of his own projects. Here, the potential for scientific

---

[9] This happens upon acceptance for publication at most journals.

misconduct is significantly higher. There are only a small number of ways that the reviewer can proceed ethically: First, he can be completely selfless and make recommendations to the author based upon his ideas. Second, after the paper has appeared in print, the reviewer can prepare a publication that demonstrates these improvements. This, however, introduces a very gray area: "When is it ethical to start this sort of research project?" It is my opinion that this sort of research *should not* start until after the original paper has been accepted for publication. At that point, most journals make the publication available to the greater scientific community. Only at this point is the reviewer not getting an unfair advantage. Does anyone actually wait this "grace period"? Probably not, and I am not aware of anyone ever being convicted of this nor am I aware of anyone getting away with it. Furthermore, it would be very difficult to get a scientific misconduct charge to stick in this sort of case.

It would certainly be unethical for the reviewer to hold up the submitted publication while the "improvements" are made and then submit these improvements for publication. Such would definitely represent gross misuse of the peer-review system. This sort of violation is also nearly impossible to catch, due to the fact that peer review is anonymous. And, although the editor knows who the reviewer is, the reviewer can lie about the delay.

Another entirely unethical scenario can be envisioned if we make the assumption that the reviewer is not working in the specific area involving the publication. If the reviewer was to then immediately start a competing project based upon this publication, he would be committing scientific misconduct. If such a reviewer intends to contribute to this field, he *must* wait until the publication has been made available to the greater scientific community. Even in this case, it is potentially very difficult to find the line between ethical and unethical. In all cases, a researcher must be very careful when he reviews a paper to not incorporate someone else's data into his own research.

Why does it happen?

The reason for this ethical violation ought to be very obvious, but a small discussion is still warranted. Obviously, if a violator is able to pass off someone else's work as his own, especially if the two researchers are competing toward the same goal, he stands to gain quite a bit. If the work could be successfully stolen from the competitor, the perpetrator would look superior.

How is it caught?

Catching this is similar to, and in some cases identical to, not acknowledging previous work done; that is, the peer-review process can often, but not always, catch the violation. Oftentimes, it can be caught if one of the reviewers is familiar with the work in question or if a particularly diligent reviewer searches for the work he is reviewing. If the reviewer is the original author of the work under threat of being stolen, such discovery would represent a high degree of poetic justice. Most often, it is only caught after the manuscript is published and the author of the original work or somebody who has read the original work happens upon the "new" manuscript.

Publication of results without consent of all the researchers

What is it?

One frequently overlooked ethical violation is that all authors must approve the publication. When it is overlooked, it is often an honest error and can be resolved peacefully. There are cases, however, that can be imagined where one of the authors does not agree with one of the chief conclusions the other authors are putting forth. In a case such as this, a resolution must be reached before the publication can ethically proceed. There are also cases where one of the authors may not even be aware of the paper; this is not unthinkable, since sometimes parts of an author's work may get published years after leaving a lab. This can occur innocently or with more sinister motives. In both cases, it is not the fault of the person who does not know about the publication. A problem can result when the publication is brought to her attention after it is in print, only then finding that she disagrees with one of the fundamental conclusions, a conclusion, for example, that is in direct opposition to one she just came to in her own lab. In a case like this, it is easy to see why that person would be a little upset that she was not consulted. This situation could also cause significant complications for the forgotten researcher when applying for a grant. In the National Science Foundation (NSF) bio-sketch, principal investigators (PIs) are required to list collaborators within a certain number of years. A situation can certainly occur, although it should be rare, where a paper is published from work completed inside this period of time yet unknown to the grant applicant. The granting agencies may then interpret this as an improperly completed bio-sketch and return the grant without review.

In gross misconduct cases, an author who may have nothing to do with an experiment is added to a paper to give it more "weight." For example, imagine adding a Nobel Laureate's name to your paper to increase its chances of getting published, and said Nobel Laureate having never even heard your name before. I have had private conversations with a reviewer for a journal where the reviewer expressed the opinion that manuscripts he or she felt were terribly written but from an eminent chemist's lab were still accepted with little or no alterations, leaving the reviewer feeling as if his or her review was ignored. The reason why it increases the chances of the paper getting published is that many reviewers and editors may be more hesitant to make negative comments on an article co-authored by a preeminent scientist, and it can certainly be argued that this is an example where peer review fails to operate as designed and may even be corrupted. In a case like this, it is the author, not the science, that is evaluated. This should clearly represent an unacceptable practice.

Why does it happen?

Publication of research without the knowledge of all authors is a violation of ethics that happens for both innocent and sinister reasons. Considering the innocent variety first, some people don't fully appreciate that it is unethical to list someone as an author for a piece of work and not run it by the other person for approval. Most PIs want the input of all the workers involved, but some people consider this an option and not an obligation. At the very least, upon submission the PI must send a copy to all the authors. However, this can occur with much more sinister motives: for example, by adding a famous researcher in the field onto the author list. This, by simple virtue of association, makes the paper a stronger paper. Whether this is "the way it should be" is irrelevant in this discussion; it happens. The likelihood of the paper being rejected or harshly criticized is immediately and drastically reduced, just by the simple inclusion of that famous researcher's name. Finally, a PI may want to keep someone in the dark about a publication because she thinks the other person might object to something in the paper. You might be tempted to say that the PI should be commended for adding such a worker as an author to begin with, but I cannot agree. You cannot make one ethical violation acceptable by going out of your way to not commit a different one. Life and science just do not work like that. Scientific misconduct is not like acquiring credits that allow you transgressions!

This violation can happen innocently as well. For example, with the high turnover of researchers (especially outstanding undergrads), it is not uncommon for researchers to lose touch. This could make contacting potential co-authors problematic, if not impossible. Social and professional networks like Facebook and LinkedIn should be harnessed to alleviate this.

To demonstrate this unfortunate tendency, consider the two cases below:

One instance where having an expert whose work is perceived as being beyond reproach may be the fraudulent claims of the discovery of element 118 by the Lawrence Berkeley National Lab in 2002.[10] In 1999, Victor Ninov, then a researcher at the Lawrence Berkeley National Laboratory, claimed to have found results that suggested the formation

---

[10] M. Jacoby, *Chemical and Engineering News*, 2002, November 4, 31–33.

of element 118. Follow-up experiments at Berkeley and other labs in Japan and Germany failed to confirm the results, and an investigation revealed Ninov had fabricated the data that suggested the formation of element 118. The paper was retracted in 2001; after an investigation, Ninov was fired. Around the time Ninov was fired from the lab at Berkeley, previous work he had performed at the Institute for Heavy Ion Research (GSI), in Darmstadt, Germany, was scrutinized. Although GSI validated the claims to elements 110 and 112 that Ninov worked on, they discovered some of the data was fabricated and did not exist in their records. When an expert claims a discovery, particularly an expert at an upper-echelon institution, few, if any, question the results critically.

Also, when fabricated results confirm our predictions, the tendency to believe the results is that much stronger. This has been pondered as a contributing factor in the case of J. Hendrick Schön of Bell Labs, who was found to have fabricated results in the area of superconductivity and molecular electronics.[11] If previous reputations did contribute to either of those works being published to begin with (and to be fair, they might not have), the point is clearly illustrated: Famous researchers are given more benefit of the doubt, and their publication road is thereby easier.

How is it caught?

This is usually caught fairly easily and handled peacefully: the uninformed author finds out, forgives the other author, and publication proceeds with all authors' approval. In other cases, the now-informed author may object to and successfully block publication. In a case like this, the issues causing the objection must be resolved before publication can move forward; simply removing the objecting author's name, even if his work is removed too, is not an acceptable option. Most times, however, this issue is resolved peacefully, occasionally with as little as some additions, removals, or corrections and, in extreme cases, an additional experiment or two being run.

Some journals are proactive about this issue, mandating that all authors give consent to the journal, not the PI. A partial list of journals that have this policy is:

- American Mathematical Society
- *Journal of Antimicrobial Chemotherapy*
    - A form that indicates all authors have complied with the stipulations in the instructions to authors is required.

[11] From *Journal of Chemical Education*, 2002, 79, 1391.

- *British Medical Journal*
  - A statement that all authors, external and internal, had full access to all of the data (including statistical reports and tables) in the study and can take responsibility for the integrity of the data; accuracy of the data analysis is required.
- Some ACS journals
  - Contact all authors directly before publication.

Failure to acknowledge all the researchers who performed the work

What is it?

This is the deliberate omission of deserving authors from the manuscript, which is quite different from the previously discussed violation. This ethical violation is subject to a great deal of gray area, perhaps even the largest amount of gray area. Journals and patents have well-articulated rules governing authorship. However, these rules are not necessarily interpreted consistently. Obviously, enormous fights could be waged over whether someone has added enough to a project (intellectually or through lab work) to warrant being a co-author on a paper or a co-inventor on a patent. With the high monetary stakes involved, especially with patents, this is an enormously important issue. It is in the best interest of the PI to be as fair as possible to all of the potential authors in these cases. Not only would an ethical violation certainly have been committed but he would gain a reputation for not granting deserving workers co-authorship that would reduce the likelihood of quality workers wanting to work for him. Potential workers would be legitimately concerned that significant contributions to a study might go unacknowledged. If a worker sees evidence that her hard work may not earn her any credit, she will inevitably search for a situation that will ensure her being given the credit she deserves for the work she does. On the other hand, it is also important to not be too generous—adding undeserving authors has the effect of diminishing each author's apparent contribution.[12] Including undeserving authors gives the appearance to the general community that it was an enormous team effort with all the authors contributing to at least one essential component of the study. An example could be a PI including as a co-author of a particular work technical staff who maintain laboratory instrumentation or provide critical insights into the interpretation of the results. While the researcher who operates the instrument or provides a critical insight should absolutely be included as a co-author, the researcher who merely performs routine maintenance should not. An exception to this rule of

---

[12] An interesting situation to consider is one where a PI requires all workers to relinquish all rights to patents. Is this ethical?

instrument operation is if you do a favor for a friend and run a sample on an instrument or a simple purification for him a single time. In this sort of case, including you as a co-author is probably not appropriate, though including your assistance in the acknowledgments probably is.[13] There is an exception, however. If this is an instrument that you designed or that your friend does not know how to operate, then you are indeed deserving of co-authorship.

Another issue that is appropriate to discuss at this point is author order. There is an enormous amount of prestige associated with being the first author on a journal article or a patent (in the United States, any-way). The first author is usually the author who contributed the most to the preparation of publication or to the work that was done. As a result, this "first authorship" is usually the most hotly contested position on a publication, though in some publications, it is the last position that is coveted.[14] Whatever the convention of the journal, author order is an issue that should be decided before the publication is prepared in the first place to keep the issue from preventing or delaying publication. In most cases, direct conversation and openness is the best approach, but determining order is not always straightforward and may require equally contributing co-workers "taking turns" being first author in cases where multiple papers are legitimately published by the same team on the same or related projects. One of the ways this has been resolved is to add a footnote that indicates certain authors contributed equally to the work.

Often, these are minor misunderstandings that are peacefully resolved. One final case is when a potential author is intentionally left off a publication due to a personal falling out or other non-science-related reasons; in this instance, a very major violation has occurred. Once again, publishers—in some cases, specific journals, professional organizations, and patent offices—have established rules regarding authorship and the reader is encouraged to consult his professional organization's or journals' rules to clarify any confusion that might arise during his career.

Why does it happen?

When deliberately committed, reduction of the author list makes it look like those remaining authors did considerably more work than they really did. In addition, it harms the omitted author for patents, because by reducing the author list a larger portion of potential royalties is left to the remaining authors. Most often, however, this violation occurs innocently. Although the PI ought to keep careful track of who does what work on

---

[13] At the very least, dinner is owed.

[14] Author orders are different in the United States than they are in Europe. Before submitting a publication, consult with editors and superiors to ensure the needs of both the publisher and the authors are met.

each project, PIs are human. They forget things, they lose track of things, and they often have several other projects with large groups of people. Reduction of the author list is often caught and handled with general peacefulness, as there is a high likelihood that the omitted author is still working in the same lab or in some way still holds a professional relationship with the PI. When it is not caught before publication of the manuscript, a corrigendum can be submitted that adds the author.

However, there can be cases where the motives are more sinister. If, for example, there is a falling out and the PI wants to get revenge on one of the workers, he may be tempted to leave the other worker off the author list of a publication. This is a clear and unfortunate ethical violation. Also, sometimes, a postdoc, graduate student, or collaborator may be bitter that the advisor or co-PI does not want to publish the results yet. If any of the other researchers attempt to publish the work and simply leave off the names of co-workers who do not want to publish yet, the result is very gross scientific misconduct. This process is somewhat tied to the violation of not obtaining permission from all of the authors. It is different here in that the objecting authors are left out altogether, though their work was not—that is, they are not added but are kept in the dark about the publication or their objections are blatantly ignored. Even if the work is repeated and the new worker is included as an author, the original author must still be included since processes he or she developed were likely used.

How is it caught?

Failing to acknowledge all authors is usually caught relatively easily. When an author sees an article that he thinks he should be listed on, he will likely make those feelings known. This is greatly facilitated by the fact that sometimes all those involved are still working together in the same laboratory. In cases where the omitted author is no longer associated with the lab or with anyone in it, however, it is much less likely that this will be caught before it is too late.

Authorship and intellectual property

The two previously mentioned infractions have other important consequences that are logical to point out now. For example, one reason it is so important to have the consent of all authors and to make sure all deserving authors are accounted for relates to intellectual property. Publication of results without consent of all the researchers can then be related to—or perhaps more accurately, inadvertently give rise to—a conflict of interest violation. Let us consider each, in turn, with hypothetical situations.

One way in which procuring the consent of all of the authors relates to intellectual property regards patents, which for some products may net the university or other institution millions of dollars. In such cases, the authors of the patent—the people who actually made the discovery or invention—almost always share a piece of that windfall. If an individual is added to a publication without his knowledge, those royalties may precipitate a conflict of interest that goes beyond the one previously mentioned with NSF bio-sketches. At the very least, such an arrangement would have to be declared at most academic institutions. If a researcher is placed onto a patent and does not know about it, the appearance of being negligent in this declaration will potentially cost that researcher her job and may result in fines or even time in prison.

Leaving deserving authors off a patent is also something that can be immensely damaging to not just one's career but one's livelihood as well. It is therefore one of the highest obligations of a PI to keep very accurate and up-to-date records of exactly which co-workers or students performed every part of every study.

Conflict of interest issues

What is it?

A conflict of interest can occur when a researcher or lab coordinator has some sort of vested (personal or financial) interest in a particular study, such that it may influence any number of things, including but not limited to, how long a failing project is continued, whether or not results are patented or published in peer-reviewed journals, or in egregious cases, even who is permitted to work on a project. A conflict of interest exists when a professor writes a book that is required for a class he teaches. Such a textbook does not immediately represent a conflict of interest. Clearly, a textbook that is written for a course by the instructor would be written in the same style and order that the material is taught by this individual. Therefore, rather than generate a true conflict of interest, an instructor-authored text has the ability to improve teaching and learning. However, it is certainly worth pointing out that a perfectly coherent argument can be put forth that argues, without using ethics and only using sound pedagogy, that this might also have the opposite effect on teaching and learning.

Conflicts of interest are perhaps best illustrated by examples. Imagine that a pharmaceutical company and an academic lab are collaborating on a study, with the pharmaceutical company providing financial support for the study, the fourth-year graduate student's salary, and a royalties-based stipend to the professor if the product ever goes to market. Assume that during the course of the study, the student discovers a new synthetic method and that this synthetic method would have impacts that reverberate through the entire field of synthetic organic chemistry. To develop this method (which would benefit the graduate student's professional career greatly) would mean that he would have to divert part of his effort away from the work with the pharmaceutical company. The research advisor might order the graduate student not to continue work on the synthetic method for fear of losing the additional royalties stipend from the pharmaceutical company. Let us be clear: As long as the appropriate institution paperwork is filed, the stipend is not unethical; the clouding of judgment it causes (if it happens) is what is unethical.

Conflicts of interest can also be present in governmental institutions such as the NSF or National Institutes of Health (NIH). If a program director for the NSF has a sibling applying for a grant in that program, a definite conflict of interest exists. Similarly, if an employee of the NIH has millions of dollars invested in a pharmaceutical company that is under investigation, it may influence her ability to do her job well.

Earlier, the topic of an instructor-authored textbook being used for a course was broached. The use of such a book is not unethical at most, if not all, institutions if the requisite paperwork has been filed. If, on the other hand, a student is required to purchase such a book and the instructor then does not teach out of the book or assign any course work from the book, a violation has almost certainly occurred. In this hypothetical case, the instructor appears to be mandating the purchase only to add to his pocket. There are mechanisms in place to protect the students from this sort of conflict of interest. One example is the University of Connecticut (UCONN), where faculty members must have a book they authored approved by a departmental committee in order to retain the royalties. In the absence of such approval, the royalties (at least a percentage of them, correlating to the number of UCONN students purchasing the book) must be donated to the university foundation or some other charity.[15]

In another case, let us assume that a research advisor has her own start-up company to market a material made in her lab. In this sort of scenario, it may be easy to fall into the trap of diverting more students' efforts toward a project that may give the professor great financial gains at the expense of the students' careers. One example of when this would have the most impact would be in the decision to author a patent instead of a journal publication. It is an important distinction that once a body of work is published, it can no longer be patented. This is important because it is then far more difficult to reap financial benefits from the work. Although in some cases this is not an issue, in cases where a company or individual investor (even the PI of the study) is supporting the research, maintaining the potential for a return on this investment is of critical importance and is certainly not unethical. In these cases, a patent disclosure would nearly always be submitted prior to publication of a traditional, journal-based manuscript. However, this becomes a problem because patents

---

[15] Personal communication, University of Connecticut president's office.

typically take much longer to prepare since a legal team or assistant must be involved to confirm the report is novel and file the requisite paperwork with federal and world patent offices. In some cases, after the work is patented, the PI and the researchers involved may sign over the rights to the work, causing potential for publication of the results to disappear or at least be greatly delayed.

Conflicts of interest can also occur in a way that helps the author of a paper or grant. On all NSF grant applications, applicants must submit a list of collaborators and affiliations along with every grant submitted. The purpose of this is not to ensure that the author/PI is treated fairly but to ensure that he is not given undue favorable treatment. It is for this reason that such relationships absolutely must be disclosed so that, for example, a grant proposal would not be reviewed by one of the authors' PhD advisors.

In a scenario that doesn't necessarily involve science, many universities offer as a benefit reduced or even remitted tuition for family members of employees, including their spouses. This inevitably causes the scenario to arise where an instructor will have in his classroom or research group his child or spouse. At small universities or in a small town, other family members (nieces, nephews, and so forth) may end up in class as well. This can be resolved a number of different ways, such as mandating or recommending (or anything in between) that the student in question takes the same course with a colleague. In cases where this is not an option, alternate provisions for the evaluation of the student work (grading) can and probably should be made. Whatever the case, it is of the utmost importance that all evaluation guidelines are clearly established and documented. The application of these guidelines should also be clearly demonstrated. It would be very wise to take advantage of modern technology and create a digital archive of all work when this potential conflict of interest is present. Extensive and accurate documentation is critical to preventing the appearance of unethical behavior when it is not occurring and to effectively defending oneself if accused.

Why does it happen?

The easiest explanation for why conflicts of interest continue to occur is that to prevent them one must go against the natural and understandable tendency to put a highest premium on one's own interests. It is no easy task to avoid this act of scientific misconduct; in fact, it may be the most difficult to avoid, as most of us do not even know when we are committing it. As an example, it would be very difficult for just about anybody to fairly evaluate a grant or publication from a competitor, who just for the sake of argument is working toward the same chemical target from a different pathway. In a case like this, the reviewing scientist must inform the editor that she has a conflict of interest and therefore must decline further involvement. This is the only ethical response, no matter how tempting it may be to hold up or reject the publication or grant to suit the reviewer's own interests. The editor of the journal or the program director of the grant could surely do a better job in picking reviewers, but this is not without its pitfalls. The primary pitfall of this approach is that the peer-review process is critically dependent on experts in the field reviewing the work under the current system. A publication or grant application will many times land in the hands of someone competing with the author. Furthermore, it is impossible for the editor to always know the specifics of what all the reviewers are working toward.

It should be noted that most journals do allow authors to request manuscripts not go to certain reviewers. This allows at least some measure of protection against this particular issue. It is exceedingly difficult for even the best of us to turn down an opportunity to gain an edge on our competitors or some (and, in particular cases, significantly more than some) extra money. When an advantage can be gained that, for example, prevents a competitor from receiving funding or delays/prevents publication of the competitor's work, some people may succumb to the temptation. Money is also at the root of the issue when things like instructor-authored texts are used in a class, financially benefiting the instructor. Familial, romantic, or sexual relationships with students or subordinates also all have clear and tangible benefits that cause the integrity of the scholarly or educational system to break down while benefiting the offender.

How is it caught?

Catching a conflict of interest is much harder than it seems at a first glance. It is not as simple as abandoning a project where a potential conflict of interest may exist at the first signs of strife. Strife is rampant in science and virtually every scientific discovery of any worth throughout all of history faced strife at some point. (Just think of those who insisted the world was round against the voices of those who knew they would fall off the planet if they sailed too far, or how many tries it took Thomas Edison to get the first lightbulb to work.) It can very easily be argued that continuing any project that faces strife is simply the good, scientific thing to do. However, sometimes a researcher may need reminding or it may need to be pointed out by a trusted colleague that the researcher's professional opinion is being overwhelmed by his personal desires or goals. In these cases, most rational people will take the opinions of their colleagues to heart. The other scenarios mentioned above are more difficult to catch for various reasons; in fact, the easiest way to catch such violations happens more or less accidentally, or if someone makes an accusation and an investigation ensues. Conflicts of interest that influence someone's grant or manuscript review are usually tough to catch. The only way to catch them is for a program director (for a grant) or an editor (for a journal) to notice that a particular reviewer rejected something and is now trying to publish something that competes with that which the reviewer rejected.

Sometimes, the best way to catch something is to put in place mechanisms that prevent them, like the book authorization process at UCONN. Various ways in which this is done, in particular at the NIH, are presented later.

Repeated publication of too-similar results

What is it?

Oftentimes, researchers try to boost their publication records by publishing multiple times on the same work or too-closely related work. This is not unlike plagiarism and in fact is a form of plagiarism called "self-plagiarism." This has the result of giving the impression that far more work is being done than actually is. One way in which it occurs is when the same project with no or minimal new results is submitted to multiple journals. It can also be argued that when a few (three to five, for the sake of argument) formerly published papers are combined into one large new paper, with little to no added insights, the same violation is at work, an argument I would agree with. There are instances, however, where breaking up work or publishing follow-up studies is absolutely appropriate.

For example, let us consider a hypothetical situation where a research group has investigated identifying different types of stars using a new type of telescope. It would be far more appropriate to include all of the results in one paper rather than fragment the research into separate publications detailing how the telescope was used to identify each type of star. That being said, if there is a type of star this new telescope can identify that others cannot, or maybe even very few others can, this would certainly be appropriate to publish alone, while incorporating the remainder of the work into a second paper. This is appropriate because it (1) reports all of the work, and (2) places appropriate emphasis on something that is novel. The same can be said about a new synthetic method that leads to a family of compounds that is subsequently found to have potent anticancer activity. Publishing these as separate papers—one that focuses on the method and the other on the biological activity—allows the appropriate focus to be placed on both breakthroughs. That being said, if it were not a new synthetic method that led to the new compounds, this would likely be unethical.

Reviews can be a sticky situation with regard to multiple publications. Annual reviews on the state of the art in a particular field are not inappropriate, provided enough new work is done over the course of one year. This, however, can be abused if an author writes multiple related reviews for different journals, even as many as two years apart. Even with how fast science moves today, it does not move quickly enough to warrant multiple reviews on the same topic each year. To be clear, it would not be unethical to write reviews for each of the different HIV inhibitors in one year. What would be unethical would be if this same author wrote a large summary review of HIV chemotherapy, using large amounts of the other, smaller reviews with little or no changes in this hypothetical larger review. This is really two violations in one. First, it is certainly repeated publication of the same results (although since it is a review, maybe data are better than results). In using the smaller reviews in the larger one, the larger one does not contribute anything novel to the scientific community. This also amounts to self-plagiarism, which is one of the reasons this is so wrong. Even if the previous smaller review is cited in the larger one, taking large amounts of work from a publication with little or no change (even from your own work) is not ethical and constitutes scientific misconduct.

Why does it happen?

Why do people repeatedly publish similar results? As mentioned earlier, such behavior increases a researcher's publication record. At most universities, especially major NIH Research Project Grant (R01) institutions, one of the most influential factors that affect the tenure decision is the publication record of the candidate. The term that is often used to describe the premium placed on publications is "publish or perish." With such stakes on the line, it is no small wonder that especially young investigators feel enormous pressure to publish and to do so at a prolific rate. Not surprisingly, this pressure eventually leads some to cut corners and begin a march toward the line of bad science vs. bad ethics—a line that some inevitably cross. Determining which weighs more heavily—the number of publications or the quality of publication—is something that varies from school to school and maybe even from committee to committee within a school.

What was said about promotion and tenure can also be said for getting grants. No granting agency will award a researcher a grant in the total absence of evidence arguing in favor of the potential for success. Some reviewers will even consider the author's perceived level of expertise in the field. Usually this expertise is measured by the number or quality (or both) of recent publications by the author in the field. Without funding, little, if any research can be done, so the extremely high emphasis on one's publication record is obvious to all in the profession. Thus, this ethical violation is often perpetrated to either improve one's job status or to even get a job.

How is it caught?

The publication of too-similar results is perhaps the easiest to catch. If when searching a topic there are multiple "versions" of the research already published, the peer-review process and a diligent reviewer should reject a paper on these grounds and a good editor will adhere to that rejection. Also, in most cases where a researcher is perpetrating this violation, he nearly always cites his own previous work.[16] Once again, a diligent reviewer will catch this simply by doing her duty as a reviewer. Another case in which this can be caught fairly easily by the peer-review process is if an author submits the same manuscript with few or no changes to multiple journals. Recall that experts in the field are the reviewers! It is common that a researcher is a reviewer for multiple journals or grants. In this sort of case, the violation is clearly identified and there is no defense. Some journals have guidelines that prohibit the submission of a rejected manuscript to particular other journals. For example, if a manuscript submitted as a brief communication to the *Journal of Organic Chemistry* is declined, the same manuscript cannot be resubmitted to the journal *Organic Letters* as a note.

When applying for a research grant from the National Science Foundation (NSF), the authors must list all other grants that are either active or under review. Although it is ordinarily meant to be a guide to measure if the proposing author will have the time to do this work, it can also be used to ensure the authors do not submit the same grant to multiple funding agencies. It should be noted here, however, that grants, unlike research reports, can be submitted to multiple places with minimal alterations to fit the specific scope of any one organization. However, some organizations (for example, the NSF) require authors to disclose all active and submitted grants. Although this is not necessarily done as a check on multiple submissions, it can function in this capacity as well.

Breach of confidentiality

What is it?

While some people look at breaches of confidentiality as "healthy competition," they couldn't be more wrong. Many major companies have their employees sign agreements that prevent them from going to the company's competitors and allowing the companies to benefit from research or "insider" knowledge. This is, of course, very similar to stealing someone's work. (Don't misunderstand: You can switch jobs and take your experience with you—you just can't take your research or other proprietary information with you!) These confidentiality agreements are signed for a reason. If we didn't hold to these ethical standards Company A could pay someone to go work for Company B for a few years and then come back

---

[16] Most researchers cite their own previous work. It is necessary and in fact more unethical to not do so, as neglecting to do so represents a failure to acknowledge previous work done.

with all of Company B's most important information. This is not "healthy competition," it is a breach of an honor code, an ethical violation, and the scientific equivalent to insider trading. That being said, if no agreements have been signed, the argument can be made that no violation has occurred. Legally speaking, this is probably indeed correct. However, the reality of the matter is that this would still represent scientific misconduct as the professional code goes above and beyond the legal code in this and other examples. Another brand of conflict of interest or breach of confidentiality may be when a reviewer does not keep information for herself but passes it on to a colleague who is competing in the same area as the author with the work under review.

Breaches of confidentiality are also discretely different from gamesmanship. To start out simply, let us use a sports-related case of gamesmanship. Football player Peyton Manning, quarterback for the Indianapolis Colts, is renowned for his ability to call plays at the line of scrimmage. Some of his opponents (especially the defenses of the Baltimore Ravens and the Tennessee Titans) have played against Manning so long that they have become familiar with the audible calls he uses. Manning has countered by not only changing what some of the audible signals mean but also employing fake or "dummy audibles" in order to keep the defense on a different page from his own. Another example from the sports world involves players who are traded and use their knowledge of the former team's signs or a former teammate's weaknesses in order to gain an advantage for his new team. Nobody would consider either of these issues a violation of some ethical code. Likewise, it is not a violation of any kind for one university to investigate how another university does something and then clearly articulate to prospective students why it employs a superior tactic to what the competing universities offer their students. Coming back to the world of science, if a former Pfizer employee informed his new place of employment, Johnson and Johnson, that Pfizer had plans to hire 100 new research and development chemists working in the heart medication unit, it would certainly not be a breach of ethical codes for Johnson and Johnson to go out of its way to make sure its best scientists in that unit were not wooed away by Pfizer. Although it may be an excessively aggressive reaction, it likewise would not be a breach of code for Johnson and Johnson to also hire new chemists for that unit, trying to keep the best and brightest away from Pfizer.

Why does it happen?

Breaches of confidentiality happen for slightly different reasons than the other cases. Usually, breaches of confidentiality occur because someone has multiple loyalties and tries to improve his standing in one of his groups. Oftentimes, there is some sort of financial attachment, which is a clear and easily understood incentive. Other times, different incentives may be presented, including material goods, advanced placement or position, or a promise of favors in the future.

How is it caught?

Breaches of confidentiality are often caught when a company or other entity notices that its competitor suddenly has something very similar to something they are investigating/producing. Of course, a researcher is allowed to change his job, leaving one company (for example, Pfizer) to go work for a competitor (for example, GlaxoSmithKline [GSK]). What would immediately set off red flags, however, would be if a specific compound or even a family of compounds being developed at Pfizer (or even investigated but subsequently rejected by Pfizer) is suddenly under development at GSK after the individual's job change. In such cases, Pfizer would certainly have grounds to make formal complaint. When this occurs, if there was indeed a breach of confidentiality and not just a commonly investigated good idea, major litigation may follow. This can especially occur if the person committing the breach signed some sort of contract forbidding the transfer of intellectual property. Communications such as e-mail may be subpoenaed and examined to prove a breach of confidentiality has happened.

Misrepresenting others' previous work

What is it?

Misrepresenting others' work is indeed a different ethical violation than deliberate fabrication of data, though it certainly can be construed as fabrication of someone else's data. One example of this violation is to present your own (and aberrant) conclusions of someone else's work as their conclusions. Let us be clear about an important point: It is certainly not inappropriate for you to come to your own conclusion about another researcher's work. What is inappropriate, however, is if you present your conclusion as the assessment of the original authors. You simply must be forthright in explaining that the conclusion you are presenting is your interpretation of the other authors' work. It would also be appropriate for you to compare your conclusion to the one presented by the original authors, providing the appropriate context.

Another more obvious form of misrepresenting others' work is when an outright fabrication has occurred, such as saying that an inhibitor was only 45 percent effective against strain A while neglecting to mention that such work also showed it was also 95 percent effective against strain B. Such a misrepresentation is clearly scientific misconduct, especially if the inhibitor you are comparing it to, one that you made, is 84 percent effective against strain A and inactive against strain B.

Why does it happen?

Although both of the hypothetical cases above achieve the same ends, the second case does so more obviously. Misrepresenting previous work done by other researchers allows you to present your work as being superior to the other researchers' work. Because such a premium is placed on arriving at the best solution to a problem, it is absolutely clear that presenting your work as better than others' is going to be beneficial.

How is it caught?

Usually, this is exceedingly difficult to catch, if it is possible at all. The only way in which this form of scientific misconduct can be caught is if the original author (or someone who is intimately familiar with his work) reads the new report and catches the "discrepancy." It can also be caught by a diligent researcher who reads both reports in an attempt to better understand the study.

Related topics that do not necessarily represent scientific misconduct are the difference between bad ethics and bad science, proving previous results incorrect (scientific progress), and the whistle-blower's dilemma. In each of the cases, the consequences can be as bad or worse than an ethical violation, though in the latter situation there are legal safeguards intended to avoid this.

Bad ethics vs. bad science

A very difficult task in many (though not all) cases is discerning between bad *ethics* and bad *science*. Before wrapping up this chapter, this issue ought to be touched upon. For this, an imaginary pair of scenarios is instructive.

Scenario 1

Frankie is an assistant professor at a major university. He is in his fourth year, and he is beginning to panic about tenure. His lab has developed a method of adding a selenium nucleophile to alkyl halides, resulting in the displacement of the halide by the selenium nucleophile. All the principles of organic chemistry suggest that if this reaction works well for alkyl halides, it ought to work even better for benzylic or allylic halides. However, when the work on these latter compounds is done in his lab, the reaction fails. Frankie omits this data and decides in the paper to say, despite this failure, "When the high efficiency of this reaction is taken into consideration along with the fact that benzylic and allylic halides are often even more efficient substrates in similar reactions, we have every reason to believe that the same will hold true in these cases as well. An investigation into this is under way."

Let us take into consideration something else that might have happened instead.

Scenario 2

Vladomir is a third-year graduate student at a major university. To have any shot at the American Chemical Society (ACS) fellowship he plans to apply for, he needs at least one more publication to be in preparation or in print. He develops a method for adding selenium nucleophiles to alkyl halides. Overrun with excitement, Vlad convinces his advisor to publish, and they publish the work they have with the intent of exploring the benzylic and allylic halides next. They make the statement: "When the high efficiency of this reaction is taken into consideration along with the fact that benzylic and allylic halides are often even more efficient substrates in similar reactions, we have every reason to believe that the same will hold true in these cases as well. Such studies will begin shortly."

In Scenario 1, it should be clear after the earlier discussion that an ethical violation has occurred. Information was known by Frankie that directly conflicted with his conclusion, and he intentionally withheld it from his publication. This falls into the category of deliberate omission of known data that does not agree with results and is an example of gross ethical misconduct. This is the case, even though the claim that an "investigation into this is underway" was made. This can surely be a defense of why no violation has occurred. Frankie can easily claim that they are indeed investigating why this did not work and that they were not convinced they had optimized the conditions yet—indeed, a plausible defense. Unfortunately, they knew they were misrepresenting the data. The only ethical way to have proceeded here would have been to present the data that they had obtained and, at the end, make a statement that sends the message: "An exploration into why the allylic and benzylic substrates failed to give superior results despite decades of research demonstrating otherwise is under way." Of course, such an investigation really must be under way!

In Scenario 2, we have assumed that the investigation into the benzylic and allylic really is under way. Without this assumption, Vlad and his mentor are behaving unethically. With this assumption, it should be equally clear that it is simply bad science. (Perhaps the case can be made that it is poor peer review in both cases, but that is a different discussion altogether and is covered in Chapter 3.) In Scenario 2, Vladomir made a pretty safe assumption and generalization. There is nothing unethical about it, even if in the long run it turns out to be a bad, even stupid, assumption. Bluntly put, it is not unethical to do something stupid! Based on his data and his previous experiences, Vlad had every reason to draw the line that he did. Unfortunately, as Frankie's data shows (data that Vlad doesn't know about), Vlad was wrong.

Let us be clear on the distinction between these two cases. What Frankie *did* was wrong, but what Vladomir *thought* was wrong. If in every case of new science we are not allowed to take for granted that certain previous trends will hold, then, really, what good is there in keeping track of anything? Making certain assumptions can be dangerous, but it is not unethical. In some cases, it comes back to burn you, but at the end of the day, assumptions are not unethical and being wrong is certainly not either. Science is largely dependent on standing on each other's shoulders. If we are not allowed to do that, work cannot progress. And finally, if Vlad being wrong *really* makes him unethical, at least he will have company— Who among us has never been wrong?

Something should also be said about the selective interpretation of data as well, as it is as much bad science as it is bad ethics. An example of this is a relatively recent report that, over a lifetime, healthy individuals had a higher lifetime health expenditure than obese people or smokers.[17] The study was done via a simulation taking various factors into account. However, the study appears to fail to consider issues like the quality of life for the individuals and also appears to overlook the fact that healthier individuals inevitably contribute more positive things to society by way of missing less time at work and working for a longer number of years. Unfortunately, such selective interpretations can result when issues such as health care become politicized and corrupted to meet the needs of the organization behind the study. The reason why such selective interpretation can be considered both bad science and bad ethics is that science must be above such taint. Science is supposed to be completely objective; when preconceived notions soil the interpretation, we are not following good science. Also, when we allow others to forcibly change our mind to suit their agenda or when we deliberately only do such experiments that prove us right and not evaluate the validity of the claim, we are employing bad ethics in addition to bad science.

New results that prove old results wrong

The above-mentioned hypothetical situations serve as an effective lead-in to a question that some readers may have at this point: "Has scientific misconduct occurred if someone is proven wrong?" The answer to this

[17] P. H. M. Van Baal, et al., *PLoS Medicine*, 2008, 5, "Lifetime Medical Costs of Obesity: Prevention No Cure for Increasing Health Expenditure."

question is, without a doubt, always "no." Science proves itself wrong on a routine basis. It is for this reason the answer to the related question, "Is it always bad science that is proven wrong?" is *almost* always "no." Instruments become more powerful and provide better resolution, allowing us, for example, to see three where we had previously seen only one. Genuine mistakes can be made in interpretation of data as well. Also, things are redefined. Twenty years ago, science said Pluto was a planet; today, science says it is not. No bad science, no bad ethics, just scientific progress. Alternatively, there may be an unidentified decomposition or other influencing factors that cause a researcher to not really be observing what he thinks he is observing. An example where yields have been improved is in the development of what we now call Grignard reagents from the Barbier coupling reaction. In short, initial results (the Barbier coupling reaction) involved mixing all the reagents together at once. The yields of these reactions were later shown by Grignard to dramatically improve if the organometallic reagent (specifically Mg) is prepared independently first. This is certainly not an example of bad ethics, nor is it an example of bad science on Barbier's part. This is simply the natural progression of science. One researcher takes the preliminary work done by another, adds his or her own new insights, and this leads to the advancement of science.

In other cases, financial obstacles are overcome and this leads to the production of more sensitive optics that, for example, allow astronomers to peer further into the universe than ever before. Does this mean that Galileo was a bad or unethical scientist for not creating these optics and using them on his first telescope? Nobody in his or her right mind would say yes. Similar cases can be mentioned ad nauseam—anything from medical technologies to the lightbulb. All are examples of the best of science.

One very common area where results are proven "wrong" is in the structure of natural products. Natural products oftentimes have enormous and complex chemical structures. Elucidating their full chemical structure is a monumental task that will certainly contain errors from time to time, even by the best researchers. Ultimately, these specific errors are often discovered when someone tries to synthesize the molecule and she finds that the synthetic sample is not identical in spectroscopic or physical properties to the authentic sample. If she is then able to alter the synthesis to furnish a different chemical entity (often only a small change, such as the direction in which a particular atom or group of atoms is pointing, is necessary) and this new chemical entity has identical spectroscopic and physical properties to the authentic sample, she has now corrected the initial structure assignment. This is something that happens—often—and it is not an indication of bad ethics nor an indication of bad science.[18] In fact, I am tempted to argue that this is actually an example of good science. It demonstrates how science checks and corrects itself without the "help" of legislation or congressional interference. It leaves everything in the hands of the people who know best—the experts.

The whistle-blower's dilemma

Unfortunately, it is all too common that the whistle-blower—that is, the individual (often a co-worker of the violator) who called attention to the violation—suffers more serious repercussions than the violator. Of course, in cases of unfounded accusations, this is the appropriate response. In other cases, however, the whistle-blower is fired or blackballed by the field. It should be made clear that peer reviewers do not usually see these repercussions. Occasionally, a researcher trying to reproduce the results may see some repercussions, although this is rare. These harsh repercussions are usually felt by those whistle-blowers with whom the violator worked. This is a most unfortunate artifact of human nature, and it contributes to unchecked scientific misconduct in all aspects of society.

Three cases (and certainly not the only three) where the whistle-blower had negative repercussions inflicted upon them are Salvador Castro, at the time a medical electronic engineer at Air Shields Inc. in Pennsylvania, and Margot O'Toole, then a postdoctoral research associate in the Imanishi-Kari lab at the Massachusetts Institute of Technology (MIT). Another case involves Suzanne Stratton of the Carle Foundation Hospital. The O'Toole case is discussed in greater detail in the context of fraud in Chapter 8, with the focus here on the effects on O'Toole herself.

We will consider Castro's case first.[19] While working at Air Shields, Inc. (at the time based in Pennsylvania) in 1995, Castro identified a serious design flaw in one of the company's infant incubators. After reporting this flaw to his supervisor produced no changes in the design, Castro threatened to file a report with the U.S. Food and Drug Administration. He was then fired. Castro sued Air Shields for wrongful termination, but the case still has not been resolved, in part because the company has changed hands more than once since firing him and since Pennsylvania employment laws at the time permitted employers to fire an employee without a reason. Searching online for updates to this lawsuit bore no fruit.

---

[18] When searching SciFinder Scholar for articles containing "structural revision" and the search was limited to articles published in 2009, thirty-eight references were found. "Structural revision" is just one of the ways in which the structure of natural products is reported in the literature.

[19] http://spectrum.ieee.org/at-work/tech-careers/the-whistleblowers-dilemma (last accessed August 24, 2011).

Regarding the O'Toole situation, many portions of the story are necessarily left out here but are discussed later in this book.[20] In short, work published by Teresa Imanishi-Kari and David Baltimore was disputed by O'Toole. It was alleged by O'Toole that Imanishi-Kari did not have data the paper claimed that she did and what was more, that the data in hand were misrepresented. At one point, after reporting her discovery, O'Toole was allegedly told by Gene Brown, then Dean of Science at MIT, that she must either make a formal charge of fraud or drop the matter entirely, an allegation Brown later denied. Only after Imanishi-Kari produced a compilation notebook during a congressional investigation did O'Toole finally allege fraud. A report from the Office of Research Integrity ultimately found Imanishi-Kari guilty, carrying a penalty of a ten-year ban from receiving federal grant money. She was also suspended from the faculty at Tufts, where she was employed at the time of the investigation. Imanishi-Kari appealed and won her appeal, though the verdict was viewed by some as "not proven" rather than "not guilty." The decision went on to criticize O'Toole, claiming she might have become too vested in the outcome of the investigation and that the extent to which she assisted the investigation might not have been appropriate. The appeals panel even questioned the accuracy of O'Toole's statements, despite her version of the events being the most consistent and unchanging of all of the persons involved in this sordid affair. She was once referred to as incoherent by one of the researchers she brought her concerns to.

Suzanne Stratton, who holds a PhD in molecular biology, was at the time of this incident the vice president of research at the Carle Foundation Hospital, a leader in cancer research. Stratton and the PI on a series of projects, Kendrith M. Rowland Jr., an oncologist, had a significant history of conflict. After two years of working at Carle, Stratton cited an outside audit that uncovered major deficiencies in twelve of twenty-nine experiments investigated that were overseen by Rowland.[21] These deficiencies had the potential to harm patients or skew results (the details are beyond the scope of this book). When Stratton voiced her concern to hospital administrators, they responded by firing her. Though they claim her complaints were not related to her firing, one is free to interpret this otherwise "amazing coincidence" however one chooses.

Wrapping up

By now, you should be noticing a theme in why scientific misconduct occurs. In just about every case, it is used by the perpetrator to gain an edge

[20] H. F. Judson, *The Great Betrayal: Fraud in Science*, Harcourt, 2004, 191–243.
[21] http://www.nytimes.com/2009/10/23/business/23carle.html?pagewanted=all (last accessed August 26, 2011).

or get ahead. This, in fact, is why all forms of cheating happen, whether it is a researcher claiming that a yield on a reaction is 89 percent when it is really 13 percent, an athlete using performance-enhancing drugs, or anything in between. Unfortunately, this is probably just human nature. We are a competitive breed and, in all likelihood, we always will be. With that assumption in mind, the best that we can probably do is improve our ability to catch instances of scientific misconduct and improve our understanding of what scientific misconduct is so that we can avoid the truly accidental lapses in scientific integrity. Science in some form has been practiced for millennia. Over that time, we have become quite proficient at cleaning up after ourselves, even without legislative help. In cases where government money was used to fund the research stricken by an ethical violation, the government certainly can and should get involved somehow. However, the international nature of science makes it very difficult for any one government to rule on ethical violations. In fact, no government should be ruling on ethical violations. This should always be left to experts. What the governments can and should do, however, is withhold federal funding from those who have been convicted by their peers.

Scientific misconduct is also significantly different from bad science. The difference must be boiled down to acting wrong and being wrong, respectively.

chapter two

What happens to those who violate responsible conduct?

The penalties for scientific misconduct vary from instance to instance. In some cases, the extent of the penalty is that the researcher's reputation is forever tainted, such as with Watson and Crick (whom some people will never forgive for the controversy that enveloped them) and Rosalind Franklin after they solved the structure of DNA.[1] In more official penalties, sanctions could be levied upon the researcher or the researcher might even be terminated.

At this point, it would probably be most instructive to discuss examples of real-life cases, starting with a case of plagiarism. In 2006, George Wagner wrote a letter to the editor of *Chemical and Engineering News*[2] describing an experience he recently had in reading an article from a group of Chinese researchers. He wrote that he found some parts of the introduction to the paper in question to be "hauntingly familiar" and, upon a brief investigation, discovered that parts of it were reprinted virtually word for word from a paper that he had previously published. Wagner promptly contacted the editor of the journal and the editor requested of the authors that an erratum and apology be written at once. An editor-in-chief of the journal that published the manuscript, *Inorganic Chemistry*, commented that "when it comes to copying boilerplate-type text for an article, 'for many Chinese authors, it's non-offensive.'" In fact, in some cultures, plagiarism is actually a show of respect and flattery. Of course, in Western culture, this is a completely inexcusable offense and furthermore is not tolerated by the science community at large. This does, however, bring us to an imbalanced set of consequences. Without doubt, there would have been at least some disciplinary action taken against a Western scientist by his institution had he committed this same offense, in addition to the mandate by the editor of the journal. No such penalties were reported here.

When fabrication of data is discovered, the penalty is very severe. The offending researcher may be fired, denied a degree (if he or she is

---

[1] M. White, *Rivals: Conflict as the Fuel of Science*, Vintage Books, 2002, 231–273; J. D. Watson, *The Double Helix: A Personal Account of the Discovery of the Structure of DNA*, Simon & Schuster, New York, 1996.

[2] http://pubs.acs.org/doi/abs/10.1021/cen-v084n025.p006 (last accessed 12/19/2011).

a student), or even have funding revoked. Further, the paper or papers containing the fabricated data would certainly be retracted so that nobody else falls victim to this crime on science, and significant embarrassment befalls the journal, the researchers, and the associated university or company.

The case of Woo Suk Hwang demonstrates the fact that not only the researcher but also the field can potentially suffer severe repercussions when fraud occurs. Hwang was found to have fabricated data that contributed to two separate papers in *Science*. His research involved cloning and human embryonic stem cells. This is noteworthy because this field of research already has shaky (at best) public support, at least in the United States. Any negative attention (and this certainly constitutes negative attention) could potentially bring funding and public support to a halt. To date, such repercussions have not happened. For his actions, Hwang nearly did time in prison. Not only was the data faked (Hwang admitted as much but claimed he was duped by a colleague), but Hwang was found guilty of fraud as well. He was found to have accepted approximately $2 billion in private funds under false pretenses and he was accused of embezzling approximately $800,000 and purchasing human eggs for research, a violation of South Korean bioethics laws. He was spared a jail sentence if he "stayed out of trouble" for three years but then was fired from Seoul National University and the South Korean government stripped him of his rights to conduct stem cell research; *Science* naturally retracted both of his disputed papers.

The story, however, indirectly continues. Protocols established in South Korea in response to this debacle were employed in 2008 to help root out another case of fraud by Tae Kook Kim and colleagues who fabricated data during their investigation of a screening technology that would allow them to identify drug targets. The investigators found that data had been both fabricated and misrepresented by the authors of the two *Science* papers. This demonstrates how far reaching, even indirectly, the repercussions of scientific misconduct can be. If the new protocols had not been in place, Kim and co-workers would have been able to get away with their fraud for a longer time. This, naturally, would have caused greater harm to science.

A short aside is important here. That both of these incidents involved the journal *Science* is not a condemnation of the quality of this journal. Science remains one of the top two or three journals in the realm of scholarly scientific publications. As a result, publication in this outstanding journal remains one of the highest, if not *the* highest, prize in scientific research. One of the outcomes of this is that work in *Science* is some of the most heavily scrutinized work, meaning that genuine errors are more likely to be uncovered than if the work were published in some obscure journal, and it makes science a bigger target than others for fabricated research. Not only is work retracted for scientific misconduct but also for reasons of good science as well. Recently, the editors-in-chief for the journals *Infection and Immunity* and *mBio* published a report described in Biotechniques.com generating a "retraction index."[3] They found that journals with a higher impact factor (such as Science) had a higher retraction index as well. Their study found that the top four retraction indices went to the New England *Journal of Medicine, Science, Cell,* and *Nature.* These four journals are four of the top five journals, with respect to impact factor, with Lancet being the fifth. The authors, while lauding the corrective nature of science, go on to point out several of the consequences of retracted articles, especially for scientific misconduct reasons that are also discussed in this book:

- Diversion of scientists down unproductive lines of research
- Unfair distribution of scientific resources
- Inappropriate medical treatment for patients
- Erosion of public confidence in science
- Erosion of public financial support for science

They also point out that the damage that even the relatively small number of retractions can do is disproportionate to the relatively low number of retractions that occur.

The editors of *Chemistry of Materials* wrote a letter to *Chemical and Engineering News* in 2005 to describe an incident that they had.[4] This case dealt with duplicate submissions and results that were too similar. Specifically, a paper in *Chemistry of Materials* was found to be essentially a duplicate of a paper in another journal. No specific details were provided by the editors in their letter but they described their decision to take fairly (and deservedly) severe action and not only withdrew the paper in question from their Web edition but also posed a one- to three-year ban on publishing in the journal on the authors. The editors go on in their letter to mention that other penalties that they would consider levying on perpetrators of scientific misconduct to include notifying their reviewers and the editors of previous journals a manuscript was found to be submitted to.

Recently a 2006 paper by former Duke associate professor Anil Potti in *Blood* was retracted after irreproducible results were found.[5] This article, though only published in 2006, had already been cited twenty-four

[3] F. C. Feng and A. Casadevall, doi:10.1128/IAI.05661–11; S. Pun, *Biotechniques.com,* September 13, 2011, "Higher Impact Factor, Higher Retraction Frequency."

[4] http://pubs.acs.org/isubscribe/journals/cen/83/i26/html/8326lett.html (last accessed December 21, 2011).

[5] A. A. Potti, H. K. Bild, D. A. Dressman, J. R. Lewis, and T. L. Ortel, *Blood,* 2006, 107, 1391–1396; M. Bialeck, *Biotechniques.com,* September 2, 2011, "Five Retractions and Counting" (last accessed September 20, 2011).

times by the time of the retraction, according to the Thomson Reuters (formerly ISI) Web of Knowledge. During an Institute of Medicine hearing, the Duke vice chancellor for clinical research, Rob Califf, testified that the university was nearly done with its investigation. Califf anticipated that of the forty papers co-authored by Potti during his tenure at Duke, more than half will be either fully retracted or partially retracted, a staggering number of retractions to say the least. Prior to the *Blood* 2006 retraction, four other Potti papers had already been retracted.[6] The investigation by Duke also discovered the Potti had lied on a grant application, claiming he was a Rhodes Scholar recipient, while he was not. Potti resigned from Duke amid the investigation while on paid administrative leave and currently holds a position at the Coastal Cancer Center in South Carolina.

It is not only academic institutions but also governmental research institutions that suffer from scientific misconduct. Starting in late 2003, the National Institutes of Health (NIH) were involved in a congressional investigation into conflicts of interest committed by institute scientists. By the end of the cooperative investigation by the NIH and a House committee, forty-four scientists were identified as having violated NIH policies or rules governing conflicts of interest. Of these, thirty-six were referred for administrative action, eight were no longer NIH employees by the end of the investigation, and nine were referred to the Department of Health and Human Services Office of Inspector General for further investigation. Additionally, the NIH identified twenty-two further cases that the House Committee had not and initiated an investigation of those scientists. (An interruption is appropriate here. In this case, the NIH and House Committee worked together productively to address serious concerns that challenged the integrity of the NIH. All of the individuals involved deserve high praise for working together harmoniously and efficiently.)

The main violation that occurred was that scientists were failing to disclose potential conflicts of interest. The transgression of these misbehaving relative few (the NIH employs thousands of scientists) brought severe repercussions down on everyone. The new rules born from the investigation can be summarized as follows:

[6] H. Bonnefoi et al., *Lancet Onocology*, 2007, 12, 1071–1078; D. S. Hsu et al., *Journal of Clinical Oncology*, 2007, 25, 4350–4357; A. Potti, H. K. Dressman, A. Bild, R. F. Riedel, G. Chang, R. Sayer, J. Cragun, H Cottrill, M. J. Kelley, R. Petersen, D. Harpole, J. Marks, A. Berchuck, G. S. Ginsburg, P. Febbo, J. Lancaster, and J. R. Nevins, *Nature Medicine*, 2006, 4, 889; A. Potti, S. Mukherjee, R. Petersen, H. K. Dressman, A. Bild, J. Koontz, R. Kratzke, M. A. Watson, M. Kelley, G. S. Ginsburg, M. West, D. H. Harpole Jr., and J. R. Nevins, *New England Journal of Medicine*, 2006, 355, 570–580.

1. NIH employees are barred from participating in paid or unpaid activities with pharmaceutical or biotech companies, health care providers, health insurers, trade and professional organizations, and higher education or research institutions that hold or are applying for agency grants.
2. Employees of the NIH are restricted from owning greater than $15,000 in pharmaceutical, biotech, or related company stocks.
3. Senior level employees of the NIH are prohibited from holding any stock in the above-mentioned sectors.
4. No employee can accept an award greater than $200 unless it is a major scientific award (e.g., Nobel Prize).

Repercussions (probably unintended ones) were felt immediately. For example, James F. Battel, at the time director of the National Institute on Deafness and Other Communication Disorders, understandably felt compelled to retire since significant family investment would have to be divested, bringing about a significant tax burden to maintain compliance with Rule 3 above. Also, a survey revealed that approximately 40 percent of tenure or tenure-track scientists were or had considered searching for a job outside NIH because of the new regulations. In late 2006, Pearson Sunderland III was sentenced to two years' probation under the new rules for activities he committed between 1998 and 2004. The maximum penalty for Sunderland's actions was up to one year in jail and a fine of $100,000. This would have been a fairly significant penalty considering that most people felt a crime had not taken place. This, however, is the importance of trust in science. If trust is violated, especially when federal dollars are influenced, the penalty must be severe to protect the trust.

Another case involves a very modern form of data manipulation. The *Journal of Biological Chemistry* has detected fraud in digitally altered photos.[7] They found cases of reuse of control images within a single publication without noting the repetition; figures from one manuscript being used in another for *new* purposes; and digital removal of contaminating bands from gel patterns. The consequences in each of these cases were the rejection or withdrawal of the papers in question and notification to the authors' institution (a common and standard penalty). Most people would immediately agree that the third of these offenses is indeed an instance of scientific misconduct. The first two, however, might not be so obvious to an inexperienced researcher. Consequently, a discussion of why these two are examples of scientific misconduct is appropriate here. First, neglecting to specifically note that a control image is being reused in a series of studies, even within the same paper, leaves out an important piece of data that puts the results into context. While it can certainly be argued that it may

---

[7] www.jbc.org/site/misc/photoshop.xhtml (last accessed April 19, 2011).

have been an honest mistake, or perhaps even bad science rather than bad ethics, these incidents must be avoided whether classified as bad science or bad ethics, especially since it is usually trivial to add a single sentence that makes known a control image's reuse. If the omission influences the interpretation of the data, it is most likely an ethical violation rather than bad science.

It might also be argued that it is logical to reuse the same figure throughout a single publication. This is a good argument, but since the matter can be resolved with the addition of a single sentence, it is better to just be clear and tell this to your readers. The second case is potentially even more confusing. A researcher is certainly allowed to mention and/ or reuse prior results, such as an image, in a subsequent paper, provided that (1) enough new research is also reported that either builds upon the research previously reported or shows it in a new light, and (2) it is properly referenced.[8] We must be careful here to notice what, specifically, the offense was. The offense was said to be that the figures from one paper were being used in another for *new* purposes. This is a very important distinction and it is this that makes this action scientific misconduct. The *Journal of Biological Chemistry*, in response to these issues, has adopted the same policy as the *Journal of Cell Biology*:

> No specific feature within an image may be enhanced, obscured, moved, removed or introduced. The groupings of images from different parts of the same gel, or from different gels, fields or exposures must be made explicit by the arrangement of the figure (e.g. using dividing lines) and in the text of the figure legend. Adjustments of brightness, contrast, or color balance are acceptable if and as long as they do not obscure or eliminate any information present in the original. Nonlinear adjustments (e.g. changes to gamma settings) must be disclosed in the figure legend.

They go on to list their procedure, which they feel will ensure the prevention/ detection of such misconduct; it concludes with: "After due process involving the JBC editors, editorial staff and ASBMB Publications Committee, papers found to contain inappropriately manipulated images will be rejected or withdrawn and the matter referred to institutional officers."

In other cases, such as the Imanishi-Kari/Baltimore case discussed in detail later in this book, a researcher may be banned from applying for

---

[8] This is not the same as using a stock file of chemical structures you prepared to increase efficiently later. Here, we are considering the reuse of a result in image form.

federally funded grants or cut out of funded grants if found guilty of scientific misconduct. As will be seen when this case is discussed later, the researcher in question was then relieved of this restriction—not because of what was considered by some to be full absolution of wrongdoing, but because of inappropriate actions on the part of the investigators. The fact remains, however, that fabrication of data is dealt with extremely severely. To ban a researcher from a family of grants is to effectively blackball him from the scientific community, as without grant funding, there is quite literally no way for a researcher to perform science. It is certainly comparable to a career death sentence.

The court of public opinion and the penalties it imposes are also tremendously powerful. For example, in early 2009, Carl Djerassi wrote a letter to *Chemical and Engineering News* objecting to a lack of thorough crediting by Trost and Dong of early Bryostatin work.[9] The action that Djerassi took great exception to was Trost and Dong's decision to cite a general review of earlier work, written by authors other than the researcher who isolated and first investigated this important class of natural products. Such issues can cause divisions in the science community. Although it can certainly be argued that a diligent reader can find the original work using the references provided by Trost and Dong, it is simply not "right." It is always most appropriate to reference the original work when crediting discoveries.

The Royal Society of Chemistry reserves the right to expel violators of responsible conduct of research from their society. In instances where publication in a particular journal, application to a particular grant, attendance at meetings, or admission into professional networks is at least in part influenced by membership in a particular society, this can be a particularly stringent (but still appropriate) penalty.

Other consequences are perhaps better explained using hypothetical situations. For example, imagine that you are visiting graduate schools that you have been accepted to and are trying to decide which to attend. During one of your visits, you speak to a student who left the group that you are most interested in joining at this university in his third year, a very risky thing to do so late in one's graduate career. After asking him why, he informs you that he left because of a dispute in authorship of a paper—he felt that he did enough work to be a co-author on the paper but one of his junior lab mates argued against his inclusion and the advisor decided not to include the third-year student as a co-author on the paper based upon the discussion with the junior lab mate. This would quite naturally give you cause for concern. It would not be unreasonable for you to rethink whether you would be best served joining this research group. This is yet another reason that one must be very careful when deciding

---

[9] C. Djerassi, *Chemical and Engineering News*, January 26, 2009.

whom to include as a co-author on a paper. Being very demanding on this important issue may discourage otherwise excellent people (in this case, you!) from working with a group, either as collaborators or as students.

Breaches of confidentiality could come along with extremely severe penalties. In mid-2010, Ke-Xue Huang, formerly of Dow AgroSciences, was arrested under the Federal Economic Espionage Act of 1996. He was accused of sending confidential information about insecticides to collaborators while authoring a legitimate review article in 2009. Huang currently awaits trial but another scientist, Liu Wen, also of Dow Chemical, was found guilty of conspiring to steal company secrets. Wen was found to have paid about $50,000 in bribes to a Dow employee to supply materials about how Dow produced certain polymers.[10]

The fallout from scientific misconduct does not have to only affect the perpetrator of the misconduct. In particular, graduate students or other laboratory associates feel the aftermath of scientific misconduct when the principle investigator (PI) is found to have committed scientific misconduct. Take, for example, the case of Elizabeth Goodwin, former faculty member at the University of Wisconsin (UW).[11] Goodwin was found to have fabricated data on a grant application. This finding was actually made by the students working in her lab. After Goodwin resigned amid the controversy, only two of the seven researchers (both of them students) working in her lab at the time were able to find other positions within the department at UW. The other four students and research specialist chose to depart UW, with one completely changing careers to enter into law school instead. The investigation continued even after Goodwin left because Bill Mellon, then associate dean of the graduate school and head of UW's research compliance, felt that the university had an obligation to investigate the charges. Part of his contention was that the misconduct involved at least two federal grant applications, a renewal and an application for new funding. Mellon went on to point out that the university must show the federal government that it is serious about the honesty of the scientists employed at UW. Curiously, there has been no question of the integrity of the data Goodwin had published in various manuscripts. It appears that the foul committed was "only" fabricating data on a grant renewal application.

This particular case is noteworthy because it was primarily a group of graduate students in the lab who discovered the fabrication committed by their research mentor. The research specialist also played a role, but the role of the graduate students should not be overshadowed by this.

---

[10] www.nytimes.com/2011/2/08/business/global/08bribe.html (last accessed August 2, 2011).

[11] http://www.uwalumni.com/home/alumniandfriends/onwisconsin/owspring2008/worms.aspx (last accessed September 22, 2011).

These graduate students were taking an enormous risk—one that might have ended their careers before they even really started. That they had the courage and sense of what was right to bring this story to light is a testament to the things that are right about science, and they should absolutely be lauded for their actions. It often takes considerable courage to even disagree with one's research mentor on a scientific topic; it is an entirely more complicated and awkward matter to take up claims of scientific misconduct against one's mentor.

With respect to repeated publications, this is one ethical violation that has a built-in penalty. As discussed in Chapter 1, more publications often mean a better career, but the rest of us in the field are not stupid. When we see multiple publications with many similar titles, we know what is happening, and yet people do it anyway. Furthermore, if one massive paper is instead divided into three or four smaller papers, the publication record is certainly thickened, but it is of great prestige to have your article cited more times. The more thinly spread the work, the less any single publications would get cited, in principle. So although it may be detrimental to the offending researcher in a more indirect way, it is certainly detrimental to the perception of the overall quality of the work in the long run.

There is one case, to date, where a researcher has been incarcerated for scientific misconduct: Eric Poehlman, a former professor at the University of Vermont (UVM).[12] Poehlman was found to have fabricated data for at least ten manuscripts submitted to different journals.[13] He also presented faked data in lectures during seminars. Poehlman also was accused of reporting false data in a funded grant application, a federal crime with a maximum penalty of five years in federal prison. During his defense, Poehlman lied under oath but did not reportedly face charges of perjury. As the case built against him, Poehlman left UVM to take a position at the University of Montreal. With pressure mounting and the threat of prison hanging over his head, Poehlman finally offered his full cooperation, ultimately pleading guilty to the charge of making fraudulent claims in a grant. He received a sentence of one year and one day in federal prison plus two years' probation. He was also ordered to pay nearly $200,000 in restitution. He was also banned *forever* from receiving public money. Poehlman's research involved searching for a correlation between weight gain and menopause, among other things. At his sentencing, Poehlman had some choice words about the scientific establishment:

> I had placed myself, in all honesty, in a situation, in
> an academic position which the amount of grants

[12] http://www.nytimes.com/2006/10/22/magazine/22sciencefraud.html?pagewanted=1&_r=1 (last accessed September 22, 2011).
[13] H. C. Sox and D. Rennie, *Annals of Internal Medicine*, 2006, 144, 609–613.

> that you held basically determined one's self-worth,
> everything flowed from that. With that grant, I
> could pay people's salaries, which I was very, very
> concerned about. I take full responsibility for the
> type of position that I had that was so grant-depen-
> dent. But it created a maladaptive behavior pattern.
> I was on a treadmill and I couldn't get off.

Such incentives for scientific misconduct are mentioned elsewhere in this book. This does not justify them, of course, nor should we feel pity on those who succumb to the pressures Eric Poehlman waxes about. The fact remains, however, that such incentives are real and they drive the foolish and unethical choices that a few (and truly just a few) scientists make.

Human and animal subjects

Regarding human and animal subjects, extreme care must be taken to ensure their proper care and oversight for their well-being. The care of human subjects is covered in more detail later. Part of why so much care must be taken is not only because life itself is at stake but also because there are severe official, financial, and federally enforced penalties and sociological penalties for the errors that can happen. Animal rights activists are among the most passionate and vocal for their cause of all groups in society. Therefore, drawing their ire is something worth avoid-ing. Although the U.S. Department of Agriculture (USDA) inspects the research facilities in this country that work with warm-blooded animals each year to ensure the humane treatment of the animals, it does not dic-tate research protocols. Instead, the agency forces the facilities to establish their own ethics oversight committees that guide their actions and deci-sions based on the Animal Welfare Act.

A recent example is the citation of two Harvard animal research labs by the USDA.[14] This facility was recently cited after the second primate death in as many years. During an investigation, Paula S. Gladue, USDA veterinary medical officer, found a number of violations of the Animal Welfare Act. These violations included unsanitary conditions in the oper-ating room, unqualified staff, and dirty animal housing facilities at the Boston campus. Regarding the unqualified staff charge, an anesthetist overdosed a nonhuman primate with an anesthetic agent during a sur-gical procedure and the animal subsequently died due to kidney fail-ure. Harvard responded to this by retraining the relevant staff in the proper use of anesthetics to try prevent such an incident in the future. Executive director of Stop Animal Exploitation Now, Michael Burdkie,

---

[14] S. Pun, *Biotechniques.com*, September 9, 2011, "Harvard Animal Research Facilities Cited."

had particularly harsh comments, saying, "When there's an incident like this, when a staff has overdosed an animal causing death, it's clear that the staff is unqualified and should not be allowed to use animals in the future." Such comments are almost surely overly harsh. It is indeed a tragic mistake but, however unfortunate, mistakes happen. Unless the worker was deliberate in his or her actions or grossly incompetent, such words are indeed too harsh. However, this is exactly the concern mentioned earlier: Animal rights activists are among the most zealous of all activists.

The room cleanliness violations stemmed from the finding that the operating room had rust and peeling paint. As if that were not enough, primate chairs were found to be covered in residue and fecal matter. The facility was given a few weeks to correct these violations to avoid further action being taken by the USDA.

Harvard's New England Primate Research Center was also found to have committed violations of the Animal Welfare Act by conducting procedures that were not approved by the facilities' Institutional Animal Care and Use committee. This may mean that one procedure was approved by the oversight committee but another nonapproved procedure was used instead. Alternatively, the researchers may have just begun work with no approval at all. Both of these scenarios are serious violations.

Previously at the New England site, the USDA found other violations as well. Specifically, a cage was cleaned in a mechanical cage washer while the cage still contained a dead nonhuman primate. It is unclear from the report why the deceased animal was left in the cage at the time of the cleaning. Following these events, Mimi Sanada, an apprentice caregiver at Jungle Friends Primate Sanctuary, responded, "These types of incidents should not happen, and a warning or a fee is not sufficient. The fact that another grave mistake was made by the same facility following this past incident is indicative of a need for a much more avid and frequent inspection to ensure all of the animals' safety." Sanada's point of this not being the first offense is particularly important. If these lapses had occurred at different research facilities across the nation, each incident would be no less tragic, but it would be less indicative of a culture problem at any one facility. That each of these incidents happened at facilities under one administrative body suggests there might be a culture that endorses a shirking of the rules.

Wrapping up

As can be seen from the previous discussions, the penalties for scientific misconduct vary significantly. There are multiple factors influencing the punishments handed down, including the severity and scope of the misconduct, the culture of the "jury" (that is, Europe vs. United States vs.

Asia), and whether federal monies were involved. This will likely always be the case, and that is probably acceptable. By and large, although socio-logical penalties may vary, the scientific ones do not. In nearly all cases (likely all, in the absence of exoneration), the guilty researchers' work will never be wholly trusted again, their fraudulent work will be retracted, and they may even find themselves blacklisted by select journals, funding agencies, or both. Many of these penalties, especially the journal blacklist-ing, is a punishment exacted by the international scientific community. Such punishments are not unfairly likened to excommunication.

chapter three

What is peer review's role in responsible conduct in research?

Although peer review is perfectly capable of detecting some blatant instances of scientific misconduct, doing so is neither trivial nor is it the primary goal of peer review. For it to be so, deceit would be assumed, rather than truth, and with that change, the whole scientific enterprise would collapse. With literally thousands of scientific journals being published every year, the sheer volume of work that currently exists (especially considering that reviewers are almost always unpaid volunteers) makes it impossible for reviewers to comment not only on the quality of the science and its presentation but also on its validity. Even determining that the submission contains novel work is not always something that can be done easily. When it is taken into account that nearly all research is carried out by experts with years of very specific research experience, it is often not practical for a reviewer to take up the task of also verifying the research, especially if the work being reviewed is a study with years of data or results from a custom-built instrument. Verification of the research is usually left to the greater scientific community after the research is published—that is, of course, if the work is important enough to be used or further developed.

Although peer review has at least an obligation to report ethical misconduct, it can certainly be argued that in some cases the peer-review process in its current incarnation may encourage limited misconduct or even be unethical itself. For example, it is not unheard of for a reviewer to reject a paper based upon one negative or aberrant data point. The impact that this has on science is that authors become hesitant to report all of their data or, worse yet, to alter their data in order to avoid having their paper rejected because of it. In cases such as these, when a reviewer is being unreasonable in his justification for rejecting a paper, the editor must take action and either opt to publish the paper despite the reviewer's objections or send the paper to an additional reviewer for what would hopefully be a more reasonable review. Do not misunderstand me: I am not blaming peer review for these forms of scientific misconduct; I am merely arguing its typical modus operandi gives the author reasons to be concerned. It is *always* the cheater's fault.

These review tactics do not go without consequences for the reviewer, of course. A contentious editor, for example, would take note of which reviewers behave in this way and avoid giving such reviewers as many papers to review in the future or at least until the reviewer proves that he has either changed his review tactics or that his action was a one-time event.

Part of why reviewers are able to get away with such behavior is that reviews for most journals are anonymous (at least from the author's point of view). This means that a reviewer need not fear retribution from the author for an unreasonably negative review. The editor certainly knows who reviewed each paper, but the editor may not know that there are personal or professional conflicts that may have influenced the review. Remember, science is based on trust. Thus, an exceedingly negative review might not appear unreasonable to the editor; it might simply appear diligent. I personally know of someone who has written extremely long and even picky reviews from both scientific and writing angles. This person is one of the most stand-up people that I know and holds himself/herself to the same standards.

Despite the difficulties with anonymous peer review, it is likely the best option. This is because it allows for exactly what was just hypothesized: a diligent reviewer to harshly criticize a paper deserving of such a critique without fear of reprisal from the authors. This check on science is essential to the progress of science. If reviewers were to be compelled to critique the manuscripts less harshly for fear of reprisals, the science would inevitably become sloppy.

The argument can also be made that anonymous reviews actually facilitate the theft of data by the reviewer. Coming back to the importance of trust in science, this argument holds no water. Even with the anonymous review system currently employed, theft like this is exceedingly rare. If it does happen, the scientific community will ensure that the guilty individual is never in position to do so again.

But can peer review prevent scientific misconduct? In some cases, the answer to this query is *absolutely* but in others, it is *absolutely not*. Either way, trying to do so is very risky. Communication in science is very much based on trust. Employing this sort of tactic assumes misconduct. The peer-review process—although it usually does not do so presently—can, for example, be modified to include an additional reviewer to check all of the cited references to at least make sure all of the references are cited properly. This may be smart anyway since it would also detect honest errors in citing as well. The peer-review process could also be modified to where all journals would require statements from all of the authors giving consent to publish, agreeing with the author order, and confirming that no other deserving authors were omitted from the manuscript. Stealing someone else's work can also easily be detected by the peer-review process in cases where the previous work done is already published. This can be done by employing an additional reviewer (perhaps even the same reviewer checking citations) to perform a "novelty evaluation" where she searches for the research topic using a combination of scholarly search tools. This also may mitigate the number of instances of intentionally neglecting previous work done, but the process is already quite good at detecting this. It might also help reveal work being submitted to different journals with too-similar results. However, if the papers were submitted simultaneously, nothing can uncover it other than the coincidental assignment of the same person to review both articles. Misrepresentation of previous work done *can* be caught, but it also can also pass unnoticed even though experts in the field serve as the reviewers; they may not always be intimately familiar enough with the cited work to know if it is being represented fairly. To expect any reviewer to read every reference (even an extra reviewer) is completely unreasonable; the reviewers are almost always unpaid volunteers, after all.

The infractions the peer review is unable to identify and perhaps should not be expected to identify are conflicts of interest and breaches of confidentiality. It can easily and logically be reasoned that the responsibility of preventing these forms of scientific misconduct lies with the institutions that the researcher in question works for. Alternatively, a reviewer can certainly commit a conflict of interest violation or a breach of confidentiality violation. In a case like this, the responsibility still does not fall on the peer-review process to detect this infraction; instead, it is the editor's job.

The infractions that peer review is categorically unequipped to detect (and arguably the worst of the infractions) are fabrication of data and the closely related deliberate omission of conflicting data. These infractions are impossible to identify via peer review. Yes, some journals like *Organic Synthesis* do check procedures, but there is no earthly way all journals could ever be set up like this. This level of fact-checking is the only way these violations can ever be found via peer review. The ironic part is that at least in the case of *Organic Synthesis*, the fact checking is not done to detect scientific misconduct. Instead, it is done to independently verify a procedure viewed as being highly important for widespread use in the synthetic community. The reality (sometimes unfortunate reality) is that these forms of scientific misconduct are discovered by the general science community—eventually. Sometimes, this happens at great expense to the researchers who try to use or replicate the fabricated results and this is, of course, unfortunate. But science is amazingly good at self correction and such violations are almost always eventually caught; in cases of influential work, they *are* always caught.

Revisiting Vlad and Frankie

Recall from Chapter 1 when we discussed the difference of bad *ethics* vs. bad *science*. Remember, Frankie deliberately ignored data he had collected and Vlad made an assumption that something he had not done (yet) would work based upon decades of data. It certainly would not have been inappropriate (although it would perhaps be draconian) for the peer reviewers to have rejected both manuscripts, insisting that the researchers perform the study on the allylic and benzylic systems and report the complete work in one paper. In this way, the peer-review process would be indirectly preventing misconduct by way of forcing the work that is being omitted (in Frankie's case) and underway (in Vlad's case) to be performed and included in the paper. Essentially, by enforcing a high standard in the science, they would inadvertently stop Frankie's misconduct.

Can peer reviewers be unethical?

Almost without a doubt, the answer to this question is "yes." First of all, we previously discussed in this chapter unfairly harsh reviews and stealing reviewed work. There is unfortunately no accountability with the current peer-review process since it is anonymous. Another instance where a peer reviewer can behave unethically is by somehow interrupting the publication process. If a reviewer is sent a publication from a competitor, and the reviewer allows the publication to sit unreviewed (or without a review being submitted) for months while his own lab finishes competing work, he is abusing the system and committing scientific misconduct. I have personally known a researcher who felt a reviewer or editor deliberately held up a manuscript so that it would not appear in the first annual issue of a high-quality journal. Why does being in the first issue matter? Major journals publicize the most cited paper in any year. That the publication of what was expected to be a high-impact manuscript was delayed to the second issue harmed this paper's citation record. If the PI's suspicion is correct, misconduct certainly occurred. The same can be said of a reviewer who makes unreasonable demands of the authors for edits or additional work before publication can proceed. It is undoubtedly the peer reviewer's prerogative to give the opinion that more work needs to be done or that serious editing needs to be done. However, if the reviewer is a serious competitor in the field and she is simultaneously working on a publication of her own, in all likelihood, she is acting with intentional malice. Fortunately, the editor can intervene and overrule such a review or even send out the manuscript to an additional reviewer if the review process is taking too long. Since the editor always knows who the reviewer is, a thoughtful editor can and will stop sending such a reviewer publications to review. Unfortunately, however, some fields are too small to allow this to be done easily.

Wrapping up

Although there are clear drawbacks to the current peer-review system, it works and it usually works exceedingly well. It can be said that making rules or laws based upon rare occurrences is foolish. It is partially for this reason that the entire peer-review process does not have to be torn down and completely renovated. The main reason, again, is the presumption of trust. Those who abuse the system and behave unethically are few and far between, despite the fact that they get enormous levels of attention when they are caught. The fact that they *are* caught is perhaps the best proof that the system works. Therefore, even with the shortcomings of the existing peer-review process, there is likely no alternative that would fix the current problems and also not create a brand new set of problems.

chapter four

What effect on the public does scientific misconduct have?

In the case of federally funded research (research that is funded by tax dollars in the United States, anyway), it can easily be argued that the researchers have an obligation to show the taxpayers what it is that they have paid for. It is exactly this argument that has led to a repository of manuscripts from federally funded research found in government publications as well as subscription journals. On the other hand, the public is comprised largely of nonscientists who have not had the training to understand and evaluate the research being reported. Consequently, in all instances, it is better to put the research through the test of peer review before "going public" with results. This allows for some level of quality control of science by scientists. In matters of extreme emergencies, the peer-review process can be sped up, but the limiting step will always be the speed of the science. Scientists and not the lay public must remain in control of funding decisions. The lay public simply does not have the experience or the education to determine the quality and the potential impact of science.

MMR and autism

One example of disastrous consequences of scientific misconduct involves the "link" between autism and vaccines. Although the study that initially demonstrated a link between the measles, mumps, and rubella (MMR) vaccine and autism has been retracted from the journal *Lancet* amid the discovery of fabricated data, it has had far-reaching impacts on society. It remains in the public psyche to some extent that the MMR vaccine causes autism and that vaccines, in general, are unsafe. People may be surprised today to hear that no such link between MMR and autism is actually confirmed to exist. The repercussions of this now discredited study have been felt for years and likely will be felt for many more. Parents have been frightened into not vaccinating their children out of a fear that it may cause autism in their child. This has the result of infectious diseases coming back, infecting and killing children who would have been vaccinated (in all likelihood, anyway) if this violation of public trust had not occurred. Thus, although this is not necessarily a case where researchers went public too soon, it perfectly demonstrates the awesome power to affect society that science possesses. This power must be handled with extreme care. There is a wonderful series of articles by Brian Deer that effectively, clearly, and fairly document this case.[1] The case is filled with bad science, fabricated data, omitted data, and alleged conflicts of interest. Anyone interested in reading about a situation that has a variety of infractions, along with penalties to the researchers and society, should read Deer's articles.

Climategate

It is impossible to not be aware of the controversy that has surrounded climate change. For the purposes of this discussion, whether climate change is real is irrelevant. What is relevant, however, is that the exposure of what has come to be known as Climategate has severely damaged the credibility of the scientists investigating and advocating climate change in the public eye. As a result, even convincing evidence of the reality of climate change is less likely to be accepted by the public and acted upon. When the importance of the science is as high as it is with global climate—that is, it affects the entire planet—trust in those who are performing and presenting the work is of the utmost importance. This violation of trust in scientists is an extreme case that demonstrates the disastrous consequences that could arise from scientific misconduct.

HIV vaccine

In a less damaging case, in September 2009, the first partially successful human immunodeficiency virus (HIV) vaccine was reported. This report was made a few weeks to a month before the study was published in the *New England Journal of Medicine*. The timing of the events suggests that the paper was already through the peer-review process at the time of the original public announcement. The researchers, in the public announcement, did not initially release the results of all of the trials done in the study. It turns out, and this was revealed after the journal published the study and the results were presented at a conference, that there were two additional analyses that argued less strongly in favor of a successful vaccine. Let us be clear on a few things:

1. No data was fabricated nor was any data embellished.
2. Data was not misrepresented, especially not to peer review.
3. Data that did not agree with the hypothesis was not omitted from peer review or withheld from the public.

[1] *BMJ*, 2011, 342, C7001; *BMJ*, 2011, 342, 5258; *BMJ*, 2011, 342, 5347.

The third one may be a little confusing. In this case, these other studies that were left out of the public eye initially merely did not provide definitive proof of a vaccine, but *did* show that it was beneficial. Only the best results were part of the public announcement. It is very difficult to label this as scientific misconduct of the same vein we have been talking about. Certainly the far more responsible thing to do would have been to report all of the results at once, even to the lay public. Given the enormous human toll that the HIV-AIDS epidemic has inflicted, *any* progress would be good news. It is therefore difficult to know what the incentives would be for neglecting full disclosure. Whatever the case, there does not appear to be any negative effects from this lapse of judgment.

Animal rights groups

The previously mentioned zeal with which animal rights activists fight for their cause may also ultimately incite violence against researchers, both those who conduct their research responsibly and those who do not. Thus, in cases where such violence was incited by unethical treatment of animals by researchers, it becomes a matter of personal safety, a very real issue of life and death in extreme cases.

Cold fusion

Perhaps the most famous of all cases involving fraud, whether alleged or confirmed, is the cold fusion incident. To be fair, no fraud has ever been confirmed, only alleged. It is entirely possible that the researchers involved suffered from bias, seeing what they wanted to see, having their scientific expertise clouded. Fusion is the process that powers stars, including our star, the sun. In theory, if this process could be harnessed in a controlled way, it could lead to a virtually limitless amount of clean energy. Unfortunately, the amount of energy that is required to cause fusion in a sustained way far exceeds the amount of energy produced by fusion with today's technology. This putatively changed when Stanley Pons and Martin Fleischmann reported in 1989 that they had performed fusion at lower temperatures and thus requiring less energy.[2] Since the publication of their work, at least one other manuscript by the pair has been withdrawn from *Nature*. Suspiciously, they refused to answer any questions. The manuscripts did not contain enough information for anyone to reproduce their results—a cardinal sin against science. Even now, more than two decades later, nobody has been able to reproduce the results Pons and Fleischmann claim to have observed. As for the effect that this has had on science and the public, it still remains. Money and time continue to be

[2] M. Fleischmann and Stanley Pons, *Journal of Electroanalytical Chemistry*, 1989, 261, 293–348.

spent on something that only two people have ever claimed to observe. If indeed they fabricated the results, the impact on society has been felt in that resources chasing down this red herring could have been better invested elsewhere. With the high stakes that currently exist with energy, this is not the time to be wasting precious resources.

Bernard Kettlewell

In some cases, even after fraud (or terribly executed science) is confirmed, the scientific establishment chooses to hold onto the validity of the experiments in question. Such was the case with Bernard Kettlewell in the 1880s.[3] Kettlewell had a theory that lighter-colored moths were easier to see by birds and thus were eaten more often, leading to a decline in the population of the lighter-colored moths. Kettlewell performed an experiment that he claimed proved that the lighter-colored moths were eaten more often, arguing that they were easier to spot, inferring natural selection at work. There were, however, several issues with Kettlewell's experiment. First, he nailed dead moths to the trees for the birds to feed on. Because the trees were dark, the lighter-colored moths inevitably were more visible to a predator. Second, the moths in question were known to rarely land and rest on tree trunks, meaning the location they were placed in during the experiment did not closely approximate real conditions. Finally, birds do not normally eat moths that are on the side of a tree. Despite these flaws in the design of the experiment, the experiments were still viewed as being valid by the scientific community. One must question why. One possible answer to this is, just like the cold fusion case mentioned above, even scientists can have their conclusions biased by what they want or expect to see. If these moths were a critical part of an ecosystem, then an experiment that argues for or against their survival or impending demise may have influenced the steps that society would have taken to protect these moths. Such influence could have far-reaching effects on society if misdirected.

Electromagnetic field and high-tension power lines

Robert Liburdy, a cell biologist at the Lawrence Berkeley National Laboratory, was a leading researcher investigating the potential dangers of the electromagnetic field (EMF). Until Liburdy began his investigation, there was no evidence that demonstrated an increased health risk due to EMF. This alone does not make it impossible for there to indeed be an increased health risk, of course. However, whenever something is reported that challenges previously held notions, the wise will look upon

[3] http://www.neatorama.com/2006/09/19/10scientific-frauds-that-rocked-the-world/ (last accessed September 22, 2011).

it with a certain degree of skepticism. Liburdy's papers claimed that the fields influenced the function of cells by disrupting calcium. It was later discovered that he had either left out, manipulated, or misrepresented data such that they agreed with his pre-experiment notion that there would be such effects. This is clearly a violation of scientific conduct and is quite different from other cases where the interpretation of the data is what was skewed by the pre-experiment notions. The deliberate alteration of data is a clear instance of scientific misconduct, of poor ethics; allowing your judgment to be influenced by your prejudices is poor science.

Fracking and pollution

Although not regarded as an instance of scientific misconduct presently, some of the controversy surrounding hydrofracking is emblematic of the influence that science can have over the public.[4] The public counts on scientists to present data regarding the safety of such processes honestly and fairly. When we fail to do so, we mislead the public in directions that it otherwise might not have followed. One example of this with respect to fracking is Conrad Volz's findings that discharge from treatment plants that accept Marcellus shale wastewater are a danger to public health. One of the treatment facilities responded by claiming that Volz was incorrect to compare the discharge water to drinking water standards. Volz, however, has been unwavering in his claims, even in the face of threats of legal action from the company. If either Volz's claims are exaggerated or if the company's claims are false, extreme damage could potentially be done to the communities near where the fracking is occurring. If Volz's (and others') claims are exaggerated or even downright false, and fracking is discontinued, jobs, money invested, and a potential relief for energy would be lost if they caused the fracking to stop. However, if the company's claims are false and the concerns that Volz points to in his study are true, then the process should be stopped immediately before irreparable damage can be done to the humans living nearby using the drinking water or to any of the other delicate ecosystems that may be affected.

Wrapping up

All scientists have an obligation to behave responsibly toward society. This not only means that they must take care to not damage the environment and not mistreat animal (or human) subjects, but also that they must not report a study to the lay public before the work has been verified. Inevitably, lay persons will act upon the report and the consequences of

---

[4] http://www.pittsburghlive.com/x/pittsburghtrib/news/pittsburgh/s_732426.html (last accessed September 22, 2011).

that could be disastrous, both from a health and safety aspect and a financial one. We therefore can only conclude that before a report of a scientific study is sent to the *New York Times* (or any other general publication), it simply must be vetted by the scientific community. There is also a subtle responsibility to science itself in that changing publicized reports of scientific breakthroughs is a surefire way to lose public trust for science.

chapter five

What constitutes responsible conduct from the point of view of human/animal subjects in research?

The U.S. Food and Drug Administration (FDA) has outlined many rules, and they can be found at the following websites or by directly contacting the FDA:

  http://www.fda.gov/ScienceResearch/SpecialTopics/Running
    ClinicalTrials/default.htm
  http://www.fda.gov/AboutFDA/CentersOffices/CDER/ucm090275.
    htm

Some of the data the FDA uses[1] to evaluate the safety of any product to be used on humans or other animals are:

- Toxicity
- Observed (demonstrated) lack of adverse side effects
- Risks of clinical studies with humans and other animals
- Any potential adverse effects, especially carcinogenic and teratogenic
- The level of use (dose and duration) that can be approved

Regarding testing on human subjects, the seminal event was the authoring of the Belmont Report, an attempt to summarize the basic ethical principles behind the National Commission for the Protection of Human Subjects of Biomedical and Behavioural Research, created in 1974 with the signing into law of the National Research Act.[2] This report made no attempt to make specific recommendations for administrative action. Instead, it attempted to provide a framework to resolve any ethical

[1] www.FDA.gov/AboutFDA/CentersOffices/CDER/ucm090275.htm (accessed July 18, 2011).
[2] http://www.brown.edu/Courses/Bio_160/Projects2000/Ethics/THEBELMONTREPORT. html (last accessed September 22, 2011); http://ohsr.od.nih.gov/guidelines/belmont.html (last accessed September 22, 2011).

problems that manifest during research that involves human beings. It is important to note that these guidelines are operative whether it is drug-based research or even sociological or survey-based research. Institutions conducting research and receiving federal funding are expected to form committees to evaluate all research projects that involve human subjects. It is the job of such committees to ensure that every possible safeguard for the participants has been taken. One of the outcomes of this was a basic set of ethical principles:

1. *Respect for persons.* Individuals must be treated as persons who can make decisions for themselves. Persons with disabilities that make them less autonomous are entitled to some sort of protection. As a result, there are two moral requirements that must be met: acknowledge autonomy and protect those with lower autonomy.

2. *Informed consent.* Persons with autonomy must be given adequate information about the study such that they can make a determination regarding their willing participation in the study. It also must be clear that the volunteer understands the information that he was given, the risks associated, and the potential benefits of the study, both to himself and for the greater good. An interesting point is that not only are people of reduced mental facilities (e.g., those suffering from Alzheimer's disease, Down syndrome) considered persons with reduced autonomy but prisoners as well. This is because prisoners may be more easily coerced into participation in a study they would not otherwise willingly participate in. Such individuals should receive extra protection to make sure they are not taken advantage of. One case of this was the use of unwilling prisoners in Nazi concentration camps during World War II. These test subjects unwillingly evaluated drugs that would then go on to help noncaptives.

3. *Beneficence.* The Hippocratic maxim "First, do no harm" applies here as well as in medical ethics. Two rules have been formulated to cover beneficence: (1) Do no harm, and (2) Maximize possible benefits and minimize possible harms. The overseeing researchers must carefully evaluate the balance between scientific progress and an individual's suffering and well-being. Nearly all medical treatments have risks associated with them. Generally speaking, all of the associated risks are even more prevalent during the years immediately after a treatment's deployment. In almost all of the cases of medical treatment that we employ today, these risks are viewed to be minimal or rare enough that the potential benefits outweighed them. It should be made absolutely clear that a patient losing his or her life or becoming permanently damaged during the course of treatment or testing is not necessarily an indication of misconduct on the part of the researchers. This, however unfortunate, is part of scientific progress.

If, on the other hand, obvious signs (or complaints) of suffering went ignored or evidence that suggested severe complications in animal testing studies (which always happen before human testing) were ignored, misconduct on the part of the researchers is obvious.

4. *Justice.* Justice in this case asks the question "Who ought to receive the benefits of research and bear its burdens?" For this, a handful of widely accepted formulations have been agreed upon. These formulations are (1) to each person an equal share; (2) to each person according to individual need; (3) to each person according to individual effort; (4) to each person according to societal contribution; and (5) to each person according to merit. For justice to be a part of the study, the study must not involve test subjects who are unlikely or unable to be among the beneficiaries of the therapy being investigated. In other words, persons in desolately poor areas cannot be used to test treatments they will be unable to afford themselves once they become available.

The report then goes on to talk about the application of the general principles to the conduct of research:

1. *Informed consent.* Consent is a process, broken into the categories of receiving information, comprehending the information, and then volunteering for the study.
   a. *Information.* Usually, the following must be made available to the subjects:
      i. The research procedure
      ii. Their purposes, risks, and anticipated benefits
      iii. Alternative procedures (when therapy is involved)
      iv. Statement offering the subject the opportunity to ask questions and to withdraw from the research, even after the research has begun

         There are studies that occasionally require that information about the research be withheld from the participants. Such is allowed if, and only if, it is clear that
         i. Incomplete disclosure is truly necessary to fulfill the goals of the research.
         ii. There are no undisclosed risks that are more severe than minimal.
         iii. There is an adequate plan for debriefing subjects when appropriate and for dissemination of research results to them.

            Under no circumstances should information about risks ever be withheld for the purposes of gaining the cooperation of the volunteer.

b. *Comprehension.* It is critical that the information be presented to the volunteers in a manner that will allow them to ask questions during the explanation and also to fully understand what the study entails. For example, it would be completely inappropriate to quickly read twenty pages of material to the volunteers and not allow them the opportunity to read it themselves. In some cases, it would be appropriate to quiz the participants before beginning the study so that they can demonstrate a clear comprehension. In cases where a participant is of limited comprehension, a third party can stand in for him and make the decision. This third party should be able to understand the subject's situation and be able to also act in his best interest. Examples of individuals with limited comprehension are infants, young children, mentally disabled patients, the terminally ill, and the comatose.

c. *Voluntariness.* Consent must be voluntarily given by the test subject. No threats or any other form of coercion can be employed to gain the volunteer's cooperation with the study. Forms of coercion that would be inappropriate would include any form of penalty for not participating in the study and excessive rewards.

2. *Assessment of risks and benefits.* One important aspect of the assessment of risks and benefits is that other modes of achieving the same ends must be put into the context of the present study.

a. The nature and scope of risks and benefits, and the systematic assessment of risks and benefits

b. The benefit-to-risk ratio must be in the favor of benefits. If the risks greatly outweigh the benefits, it would be improper to conduct the study, and the committee overseeing such work at an institution would not give the researcher approval.

3. *Selection of subjects.* With the exception of studies that seek to investigate a particular group, selection of the subjects cannot be made based upon social, racial, sexual, and cultural biases that persist in society. Furthermore, selection cannot be done on a basis that allows only specific groups to benefit.

There are various issues, in addition to a strict adherence to the federal regulatory process outlined above, that are worth discussing. First, the participants in the study must be afforded full disclosure of the hazards associated with, for example, the drugs during earlier testing. If the earlier phase trials demonstrated that six out of ten mice suffered strokes at high doses of the drug, the human test subjects *must* be told this. They likewise must be told how efficacious the drug was in earlier animal testing. To be sure, the participant is much more "science experiment" than patient, but full disclosure is imperative since it *will* influence someone's willingness to be a science experiment. Undoubtedly, some people volunteer for such trials as a last chance at surviving the disease they are afflicted with. More altruistic volunteers do so to help humankind. Undoubtedly, there are also some who do it for the love of science. Also, subjects must be permitted to back out of the study at any time.

Everyone knows that some drugs have severe side effects. Some of these may involve varying degrees of pain or discomfort. Different people respond to pain differently, making pain something that is phenomenally difficult to measure. For example, somebody not accustomed to gastrointestinal illness may find a drug that gives them diarrhea for an entire week to be inhumane punishment. Furthermore, somebody may simply have a very low pain tolerance, making a drug that causes joint pain inhumane to her. This makes measuring the pain or suffering very difficult to approach ethically. That is too say, how much is too much? At what point would a participant be removed from the study to reduce his pain and suffering if he does not remove himself? How much pain must a tester experience before it has to be reported as a "rare side effect"? Also, some drugs are used despite very high toxicity. A perfect example of this is the anti-HIV drug AZT, once one of the only treatments for this disease. AZT is known to have bone marrow toxicity. Sometimes, if a drug is one of the precious few that has efficacy against a death sentence like HIV, pancreatic cancer, or inoperable brain cancer, even severe side effects may be overlooked to give the patient a chance. As long as the patient is not kept in the dark about these risks, nothing unethical has occurred.

With regard to animal testing, pain and some other side effects (such as lucid dreams) become nearly impossible to measure accurately. Most people are observant and aware enough to be able to look at an animal and know it is in discomfort. What is not always immediately clear, however, is determining what is wrong. Something like a limp is obvious enough, as is diarrhea, but a dog cannot say, "I have a headache"; a cat cannot say, "My throat hurts"; a horse cannot say, "My stomach is killing me"; and hamsters do not lament, "You would not believe the dream I had." The list can go on quite easily. Therefore, such negative side effects, and especially their severity, become almost impossible to measure during animal testing, making the humans "guinea pigs."

As discussed earlier, the U.S. Department of Agriculture (USDA) does not dictate research protocols; it forces the facilities to establish their own ethics oversight committees that guide their actions and decisions based on the Animal Welfare Act. The Animal Welfare Act sets the following guidelines:[3]

[3] http://www.aphis.usda.gov/publications/animal_welfare/content/printable_version/ fs_awawact.pdf (last accessed September 20, 2011).

1. Adequate care and treatment for housing, handling, sanitation, nutrition, water, veterinary care, and extreme conditions must be provided by the facilities.
2. Dogs must be provided opportunities to exercise.
3. Primates must be provided opportunity for psychological well-being.
4. Anesthesia or pain-relieving medication must be provided to minimize pain or distress.
5. Unnecessary duplication of specific experiments using regulated animals is prohibited.
6. An institutional animal care and use committee that will oversee the use of animals in the experiments must be established.
   a. This committee is then responsible for ensuring the facility complies with the Animal Welfare Act and providing documentation of compliance with the Animal and Plant Health Inspection Service.
   b. The committee must contain at least three members, and membership must include one veterinarian and one person not affiliated with the facility.

Another important question with regard to animal testing is, "Which animals count?" The reality is that some animals count more than others. For example, cold-blooded animals are exempt from coverage. Specifically, vertebrates count more than invertebrates, and mice are near the bottom of the totem pole of vertebrates. Vertebrate animals (even mice) require vigorous prior approval and careful monitoring, while invertebrates such as insects require no approvals or monitoring. Some people, without a doubt, will have a moral problem with animal testing, and this is their right. A former co-worker of mine once seriously stated that animal testing should never be done, preferring to just test things on people. The fact of the matter remains that there are well-established and agreed-upon rules that, when followed, avoid a form of scientific misconduct that is not scientific misconduct in the same way that was discussed earlier.

In a related topic to drug testing, a close examination of some drug development literature will reveal something that the lay reader may interpret as grotesquely unethical. Sometimes the drug that is pursued is not the one with the greatest in vitro activity and even not the highest in vivo efficacy.[4] There are several perfectly good scientific and ethical reasons why this would be the case. Firstly, a closer reading sometimes will reveal that the more active drug has significant toxicities associated with it as well. This is certainly possible and very unfortunate. In this sort of case, the less toxic drug is the only viable option to avoid this more

---

[4] In vitro refers to assay studies done on cell cultures, while in vivo refers to studies done in living organisms.

active drug's side effect. Secondly, there will also be cases where a less active drug will ironically be active against a wider range of strains of a disease. As an example, a drug may have record levels of activity against a single strain of HIV, but minimal activity against many others. Compare this to a less active drug, one that is roughly equi-active against all strains of HIV. In this case, the less active drug would logically be the one pursued since it would have the most widespread use. Finally, and this last instance would probably not be immediately apparent in the literature, there may be synthetic difficulties that dramatically increase the costs of drug production. You may be tempted to say "So what?" since, after all, it is the company's job to produce these drugs. Although this is not necessarily incorrect or unfair, per se, it is the company's prerogative to make a profit. The more money it costs to produce large quantities of the drug, the more money the treatment will cost. With this in mind, synthetic difficulties that triple the cost of production are certainly viable reasons to avoid a drug candidate as production costs would drastically increase the price of the drug. When one recognizes that some diseases—for example, malaria and HIV—are most devastating to people who are poor beyond comprehension, the need for low-cost medication (even with the assistance of philanthropists like Bill and Melinda Gates) is obvious.

One other case is when there are two candidates—we will call them drug A and drug B—that are being pursued as anti-HIV treatments. Let us assume that drug B is five times more active than drug A. Let us also assume that drug A (the less active one) is 15 percent more active than every other drug on the market but one, and this one has complications that neither drug A nor drug B suffers from. In this sort of hypothetical case, drug A may (and perhaps *should*) be released first because after resistance to drug A is acquired by the virus (which in HIV's specific case is almost inevitable), drug B can then (at least potentially) be deployed to replace it. This sort of tactic usually does not work in the reverse direction. Similar cases are seen with methicillin-resistant Staphylococcus aureus (MRSA). Cases of severe bacterial infections are increasingly in the public eye. There are some antibiotics that combat some of these infections. When an infection is present in a patient, the antibiotics are used in a specific order. This is not unethical; it is simply smart science and medicine. Immediately deploying the antibiotics that represent the last line of defense will only breed strains resistant to these antibiotics as well, resulting in a complete catastrophe. In any case, all of the data from the drug candidates must always be presented, especially in patent applications, even if the candidate drug will not be used.

Testing involving human subjects is guided by the findings in the Belmont report.

Wrapping up

Whatever one's feelings are toward drug testing, whether it is human or animal testing, such analysis is necessary. Without testing, we would have no ability to identify which drugs work and which ones do not. For this reason, the Food and Drug Administration (FDA) and its counterparts in other countries have very rigorous regulations that govern the process. When the system is followed, it works very well and minimizes the suffering that may be associated with the testing of drugs. Furthermore, drugs must be continually monitored for safety, even after deployment, and we must take care to not lose some of our best hopes for recovery to disease resistance.

chapter six

Can intervention or interference by the federal government result in research misconduct?

Embargoes enforced against other nations have caused some scientific controversy. For example, the U.S. Treasury Department's Office of Foreign Affairs Control (OFAC) has caused the delay or prevention of some publications in recent years. Furthermore, the American Chemical Society (ACS) and other American publishers have been prohibited from providing comments via the peer-review process or any other editorial services to authors from Cuba, Iran, Iraq, Libya, North Korea, and Sudan in the past. The OFAC ruling further stated that manuscripts accepted by a journal may only be reproduced in exactly the form that they are received. Let us be clear about what that says: It says that a journal, if it is going to accept a publication from an Iranian scientist, must accept it with any and all typographical errors and in the layout (which is different from the final, print layout) it is received in. This ruling included peer-review activities that would provide scientific feedback as well. This was eventually ignored by at least the *Journal of Organic Chemistry*, as indicated in a published editorial, while the ACS worked to get the ruling overturned. In late 2004, the regulations were eased after a lawsuit. In essence, the federal government has tried to dictate to an independent publisher how and what to publish. This can absolutely be argued to be a usurpation of constitutional rights. However, a counterargument that constitutional rights are extended to individuals and not organizations is certainly one that is logical. That being said, an environment where federal law inhibits the progress of science by forbidding productive communication between researchers is somewhat akin to the Dark Ages when the Catholic Church stifled science.

As if this were not enough, in late 2006 the ACS temporarily expelled thirty-six of its Iranian members along with one Sudanese member from its ranks amid concerns of violating OFAC embargos administered by the Treasury Department. About a month later, the ACS reinstated these members with several restrictions. It was the ACS's expressed intent to lift these restrictions upon receipt of a license from OFAC and this has subsequently taken place. These expulsions were not a minor public relations problem for the American Chemical Society. It ignited a month-long "blogfest" and was mentioned (at its merciful conclusion) in a June 2007 issue of *Physics Today*.

However, are the government regulations that caused these issues examples of scientific misconduct, or do they *force* misconduct? Perhaps it is neither, but it certainly strikes at one of and perhaps the most critical foundation of modern scientific enterprise: free and honest discussion and collaboration. Regulations such as these not only hurt science by inhibiting or preventing discussion but also by preventing the scientists in our own country from benefiting or using truly impressive work that may be done in other countries.

In August 2007, the U.S. Patent and Trademark Office (PTO) passed new regulations that were drawn up by the Bush administration. These rules were immediately opposed by several companies, including GlaxoSmithKline (GSK). Because of these collective objections, the rules were put on hold while the two sides argued in court. In short, the rules would have limited the number of claims that could be made in an individual patent filing. The goal of these rules was to help the PTO reduce the backlog of unexamined patents and reduce the length of time it ordinarily takes to complete the application review. It is unclear that this change would have reached these ends due to the fact that, in all likelihood, the number of applications would have increased under these rules, perhaps only making the matter worse. GSK argued, and this argument is indeed very sound, that this would have the effect of stifling innovation due to the fact that the applications are often constantly evolving on a drug candidate's pathway toward becoming a drug. Also, although this point was not reported to be part of GSK's argument, these rules would have potentially caused an application to be less thorough and, consequently, less likely to completely protect the interests of the filer. Although the rules were rescinded in 2009, the entire controversy represents an unfortunate foray into science by the government and how, even when the government has truly positive intentions, its involvement can cause significant harm. When regulations governing science are authored by legislators without extensive input from scientists, this becomes a certainty.

The issue of funding controversial research, such as stem cell research, is also one that unfortunately the government cannot help but be entangled. In this case, it is certainly not the government's fault. Those who oppose stem cell research usually do so from a religious perspective, to the great anger of those who support it. The supporters claim that such grounds for opposition are a violation of church and state separation guaranteed by the First Amendment of the Constitution. This oft misunderstood clause in the Constitution does not in any way (in my opinion) prohibit a governmental agency from allowing funding decisions to be influenced by religious convictions. The relevant portion of the First

Amendment reads: "Congress shall make no law respecting an establishment of religion, or prohibiting the free expression thereof." When it is
read with the lens of the late 1700s, this means that Congress shall not
make a law that makes the following of any particular religion illegal and
that Congress cannot create a religion. In the Constitution, the authors
in all likelihood meant establishment as a verb and not a noun, though
modern reading treats it like a noun when invoking separation of church
and state. One must keep in mind that many of those who originally came
to what were then the British colonies came here to escape religious per-
secution, including and perhaps especially from the present-day Church
of England, formed during the second schism from the Catholic Church
in 1570.

   Although not related to our federal government and also not necessarily related to misconduct, other governments have also caused quite a
stir. For example, in the past, German authorities did not allow scholars
educated in the United States to use the title "Dr." without special permission.[1] Today, we look at such rules as being arcane. And, indeed, this
controversy has been quashed. One of the cases that attracted a good deal
of attention was that of Ian Baldwin in 2008. Dr. Baldwin received a sum-
mons from the city of Jena's Criminal Investigation Department claiming
he was a suspect in an investigation into the abuse of a title. Baldwin faced
a fine and up to one year in prison for violation of a law—with its origins
in 1939—that states that only holders of a doctorate from EU nations were
permitted to use the title "Dr." American-educated PhDs were not permitted to use this title. The directors of the Max Planck Institute (Baldwin's
employer) retained legal counsel and fought the law. Since the eruption
of this controversy, state education ministers have met in Berlin and have
decided that American PhDs are now legally permitted to use the title
"Dr." in Germany.

Wrapping up

Although the government's intentions are hopefully always good, it often
does nothing but make matters worse when the government gets involved
in science (the National Institutes of Health, National Science Foundation,
and National Aeronautics and Space Administration notwithstanding
though—these federal entities are run by scientists, not lawmakers). The
clearest case of this is the OFAC rulings previously discussed. Their intent
was to prevent countries belligerent to the United States from benefiting
from services provided by the United States. The effect that they instead
produced was one that challenged one of the very essences of science:

[1] http://www.spiegel.de/international/germany/0,1518,540459,00.html   (last   accessed
   September 20, 2011).

uncensored and unrestricted communication between scientists. The one case where federal intervention is appropriate is in matters of federally funded research tainted by scientific misconduct. Even this, however, comes with a stipulation: Scientists and not lawmakers should make up the "jury" in the rulings of fraud. From there, the federal government should be allowed to impose any legal penalties it sees fit on the offending scientists in addition to any of the penalties the scientific community imposes.

chapter seven

Can we prevent misconduct in research?

Unfortunately, in the absence of some sort of worldwide change in human nature, ethical violations are probably not completely preventable. Terribly draconian penalties such as revocation of tenure, immediate termination, or even being sentenced to a multiyear publication or grant submission ban on the offending author may help, but the potential for effectiveness is at best questionable. However, it is very likely that through proper education, the number of cases can be reduced—a major goal of this book. Such a reduction can be effected at least two different ways: (1) What is now understood to be an ethical violation had not previously been known to be and will now not take place. (2) There will now exist a heightened awareness such that a potential accidental perpetrator will catch himself or herself before committing an ethical violation. Additionally, all institutions have some form of an Office of Research Integrity whose job it is to verify that its scientists are not committing scientific misconduct. Such offices are usually where investigations begin when charges of fraud are made.

One particularly severe penalty may be to mandate that researchers who commit fraud of any kind would have to pay back a percentage of any grants. Presumably, the percentage would be determined by the severity of the fraud and the number of years left on the grant, or perhaps repayment of the entire grant if the grant is over and had been funded based ultimately on fabricated work. Of course, it is highly debatable whether or not such penalties would be effective deterrents. A prime example of the potential for extremely draconian penalties is the death penalty, which has debatable effectiveness at preventing capital crime. Nobody would suggest that researchers who commit fraud should be put to death—however, if death is not an effective method of deterring capital crime, there is no reason to suppose that paying some percentage (or all) of a grant back would be effective at preventing fraud.

Intentional negligence in acknowledgment of previous work

The only really effective way to prevent or reduce this violation of proper scientific conduct is via the peer-review process where it can potentially be caught and resolved before publication. With diligent peer reviewers, work that is neglected can be included in the manuscript. Of course, the editor's cooperation is essential to this success. If a peer reviewer thinks that something should be referenced, the editor should take her recommendation into consideration. Safeguards must be put in place, however, that prevent reviewers from abusing this for their own benefit. Even with anonymous peer reviews, the editor always knows who the reviewers are. Since reviewers are ordinarily experts in a field, they may occasionally review a paper that ought to acknowledge some of their own work in the area. However, if a reviewer makes a chronic habit of making such recommendations, the editor should put an end to this and either not use the review or ask for an additional review from a perhaps less-biased reviewer.

Deliberate fabrication of data

Wholesale prevention of data fabrication is likely impossible. Fortunately, it is almost always found out, at least eventually. Although there is nothing to prevent a researcher from simply claiming mistakes were made, especially if the fabricated data is written right into a lab notebook (the first place an investigator would look). One tactic that may reduce the incidence of data fabrication would be to remove or reduce the incentives (that is, awards) for great scientific achievement. This would be completely unreasonable because, under this type of system, the greatest minds would have no clear motivation to enter into a scientific career. The only other tactic that might be effective would be to increase the penalties for verified instances of fabrication of data, exposing the offending researchers to penalties such as lawsuits charged by people directly harmed by the fabricated data. Once again, however, the effectiveness of such penalties is by no means ensured. One immediate complication that arises under such a scheme, however, would be determining just who is responsible. For example, if a student does a masterful job at deceiving her advisor, who should pay the price? The professor, who ultimately is responsible for the quality and integrity of the work? Or is it always the person who actually committed the fraud? (Suing a graduate student may be the worst way to get money ever conceived, however.)

Deliberate omission of known data that doesn't agree with hypotheses

Omitting data that does not agree with the hypothesis is quite different from fabrication of data and quite easy to address. This transgression may be borderline defensible. There is a way to prevent this ethical violation from occurring. Previously we supposed this usually occurs because aberrant data is often cause for rejecting a paper via the peer-review process. This is a practice that simply must be ended in my opinion. Work is more complete when the limits are presented and to reject work that discusses those limits is unscientific in every way. Part of the essence of science is knowing not only what works well but also what does not work as well. That way, the all-important "why?" can be explored and perhaps the limitations can even be overcome. Efforts simply must be made to change the paradigm away from this type of review. If the consequences of incorporating such data were removed, researchers will probably be less likely to omit data. Perhaps it would be most appropriate to begin this culture earlier in a student's education, even at the high school level, putting larger premiums on explaining the results they observe during a laboratory exercise. It may be worth it to deliberately install experiments with mediocre (at best) success to get students "used to" incorporating such results. That being said, we cannot have a wholesale exodus from caring about positive results. A student who gets a 13 percent yield on a chemical reaction while the rest of the class gets greater than 90 percent must receive some sort of grade penalty, no matter how logical and eloquently articulated the explanation is.

Passing another researcher's data as one's own

Similar to the fabrication of data, this can likely not be truly prevented; however, its reduction warrants an expanded discussion. It may be possible to reduce this violation through education. For example, it really is not "competition" and "scooping" your rival researcher if you take their work, quickly reproduce it so you can say that *you* actually did it, and then try to publish it as your own. This is simply not right because the work was usually presented to you in good faith as a private communication in a department seminar or in an article or grant you are reviewing. With more education, perhaps more researchers can learn that this is an unacceptable practice in every way; it is not good healthy competition. Again, to be fair, this exact type of instance is very rare. Likewise, it is against the norms of science to present previous work done by either your lab or others and not accurately and properly cite this work. By increasing awareness of self-plagiarism, this brand of scientific misconduct can be reduced.

Other than increasing education and awareness of this violation of scientific code and trust, or extremely draconian penalties, the best way to prevent this may be to increase the number of reviewers, thereby increasing the likelihood that someone will notice. This, however, is fraught with impracticalities. As stated before, peer reviewers are nearly all unpaid volunteers. Asking more reviewers to give up already precious little time is unreasonable. Furthermore, once again, it is not the intended job of peer review to ferret out these violations.

Publication of results without consent of all the researchers

Preventing this is truly quite easy. Every author should be sent a copy of the paper prior to submission and given a deadline by which to read it (at least two weeks). This provides enough time to read the paper and voice any objections or have there be consent by way of not raising objections. Some journals require a statement from each author while others contact all of the authors. All journals should adopt these sorts of practices. Another possible route would be to have all publications be submitted to journals through the office of sponsored research, the same office that often handles grant submissions at academic institutions. With this sort of administrative oversight, these issues could conceivably be more easily addressed prior to the submission of a manuscript, when there is still time to resolve them easily. There are at least two issues, however, that make this an unworkable proposal. First, such offices are not set up to handle the workload that would result from this. Second, collaborations are often researchers at different institutions. The question that would then inevitably come up is "Whose office handles this?"

Failure to acknowledge all the researchers who performed the work

In cases where someone is innocently left off a paper, the only way to ensure its prevention seems to be to encourage lab directors to keep better records and simply be more diligent. None of us are perfect, however, so eliminating this violation through this method is likely a long shot. However, social and professional networks online seem logical places to start. In cases where it is intentional, due to a personal conflict, there may be no preventing it. Human nature causes us to seek revenge, and this is one method of professional revenge in science. In other cases, where authorship is contested or argued, it would be most helpful and logical to discuss at the outset of a project or at least prior to preparing a manuscript what the parameters for authorship are. This is a time-consuming endeavor, but it would be very effective at reducing controversies.

To resolve this, and the previous violation, at least with respect to patents, all patent applications are completed with the assistance of a lawyer. This provides an ideal mechanism by which all authors can be included rightfully. Indeed, all patent applications require the permanent address and signature of all of the authors/inventors. Of course, this only includes the authors the principle investigator (PI) lists as co-inventors on the patent. There is, however, something that can be done. The attorney can (and perhaps should) interview *all* of the workers in the labs involved in the patent to ensure that no one who feels that he or she contributed has been left off the patent. If someone charges that he feels this way, there is time to investigate and resolve such claims before the patent is ultimately filed. This may cost a fair amount of money for the increased time demanded of the counsel. However, since litigation that accompanies controversy around patent authorship is always significantly more expensive than this, it may be worth the investment.

Similar regulations can be put into place regarding journal publications as well, though certainly not with lawyers. A university or other institutional office, such as the office of sponsored research or some intradepartmental committee, could in principle at least require a statement from all of the members of the research group attesting to the author list on every publication. At the very least, the PI could create a paper trail that documents that nobody in the research group feels he or she has been left out. Such a process would be workable for even the most prolifically productive research groups and institutions.

Although not formally a preventative measure per se, something can be done if this issue is uncovered after a paper is in publication. The same can be said for author order disputes. A corrigendum (or correction) could be filed and published that fixes the wrong credit.

Conflict-of-interest issues

The fast and draconian way to fix conflict-of-interest issues is to mandate that no academic lab directors are allowed to have any financial interest in their work. Clearly, this is an inappropriate solution because if a lab or researcher develops something that makes, say, a pharmaceutical company millions of dollars, the academicians deserve something for their role in the discovery beyond a pat on the back. This sort of mandate would also undoubtedly have negative repercussions on books. Most academic texts are written by professors and placing oppressive restrictions on their production would greatly harm academic progress. Faculty persons should also be allowed to use their own personal money to start up a company based on their research to fund and subsequently profit from their research. The best thing to do is have the university or other administrating body hire an independent mediator when such conflicts of interests occur or, failing that, the researcher can ask a trusted colleague to point out when his personal judgment is superseding his professional judgment. Most universities, in fact, do have similar processes as well as requirements for rigorous records that contain declarations of conflicts of interest for monitoring and self-auditing purposes. These are often enough to curb instances of conflicts of interest since they declare specifically that they may exist, resulting in everyone watching.

Repeated publication of too-similar results

This is one that the peer reviewers and editors must team up to address. With reviewers and editors across journals working together, duplicate manuscripts would be rejected and prevented. Complete prevention is unlikely and maybe even be impossible due to the volume of work being done, but this can at the very least be reduced. One potential solution may be to select a reviewer whose sole task is to check if any similar work has previously been published. This can be done quite simply with a handful of searches on SciFinder Scholar or a similar database. However, once again, this is not the intent of peer review and may not be an appropriate use of resources as a result.

Breach of confidentiality

Like other violations mentioned above, the only likely way to prevent or reduce a breach of confidentiality is to make the penalties as harsh as possible. Once again, however, the potential for success is questionable. An alternative may be to reduce the financial incentive to commit this breach for the offender, but there is almost always somebody willing to pay more for information, so that is likely to be an unsuccessful tactic.

Misrepresenting others' work

Like fabrication of data, preventing this is probably not possible. The best we could hope for is to catch the violators sooner and since, like other examples, it involves increasing the demands on peer review, it is unlikely to be implemented. There is a chance that if peer review were to stop rejecting papers for bad results, one of the incentives to do this would be reduced, but this is probably a long shot since there are multiple incentives for this violation to occur.

Wrapping up

Unfortunately, wholesale prevention of scientific misconduct is simply never going to happen. The best that we can probably reasonably expect is catch it more quickly. In all but the cases that involve fabricated or omitted data, this can likely be done—although doing so would be an enormous order. More likely, however, there will always be bad apples who will try to beat the system. Failing lifetime bans for even the smallest misconducts (which may not be effective), scientific misconduct will probably continue in some form or another.

chapter eight

Case Studies

Unlike most of the examples covered previously in this book, for which it was clear that misconduct occurred and penalties were levied, some of the cases that follow in this section are much less clear-cut. It is the intent behind this presentation that these cases stimulate conversation (and perhaps even lively debate) about whether misconduct has taken place.

Darwin and Wallace

Summary

Charles Darwin, credited with the discovery of evolution, was actually not alone in this discovery. His contemporary, Alfred Wallace, is forgotten to all but a learned circle. Their papers, presenting their findings, were jointly presented at the Linnean Society in Burlington House, Piccadilly. That Wallace's contribution is so forgotten is evident in everyday colloquialisms such as Darwinism, Social Darwinism, and the iconic Darwin Awards.

What happened?

It would not be a stretch to say that Wallace did essentially what Darwin did, only in Brazil (and thus, logically, with other specific animals). One of the major differences in their interpretations of the results is that while Darwin pursued natural selection as the impetus for evolution, Wallace pursued environmental forces as its incentive. Looking at these two explanations with modern eyes reveals little difference between the two, since the fittest survive in their own specific environment. But during their time, these were different. Furthermore, Wallace performed his studies nearly two decades after Darwin performed his. To be fair, Wallace suffered a major setback when nearly everything was lost in a ship fire, contributing to his delay. So what happened that their studies were released together? One must keep in mind that this was the mid to early 1800s. The feelings about creationism then were far more intense than they are, even today. In that time (in Judeo-Christian societies anyway), belief in anything but the biblical view was literally considered heresy. With the Catholic Church (especially in Europe) wielding so much power, being branded a heretic then was one of the worst sentences one could incur. Therefore, Darwin's hesitation and his preference to be absolutely sure are quite understandable.

Darwin received the confidence he needed from a letter sent to him by Wallace (who also shared bird samples with Darwin) in which Wallace described his theory. This theory allegedly came to Wallace in a fever-induced dream. His letter was not the first time the two had communicated. He sent this letter to Darwin requesting that Darwin review the paper and if he (Darwin) thought it worthy, to pass it on to Charles Lyell. To Darwin's horror, this theory was very similar to his own. What happened next was, at the very least, curious. Darwin wrote to friends Joseph Hooker and Charles Lyell lamenting that someone might get the credit for his discovery. It should be noted that Darwin was a far more respected and renowned researcher than Wallace. Had the theory only been presented by Wallace, it might have done more damage to the theory than good. Hooker and Lyell determined (and their friend Darwin's interests were almost undoubtedly an incentive for them) to present both theories together.

Wallace, for his part, was allegedly pleased with this outcome. It gave him significant credit (even if some of it is lost in the translation to modern times) and effectively inducted him into an "inner circle" of sorts containing the greatest and most respected scientific minds of the day. He later went on to more or less father the field of biogeography. Meanwhile, Darwin proceeded to author his epic *Origin of Species*. Perhaps a combination of these last two points is the real reason Darwin is the receptor of most of the popular credit. Darwin continued contributing to the field while Wallace pursued another, albeit related, one. Wallace did not even once publically cry foul against Darwin, Lyell, or Hooker. That has not stopped some from writing treatises about Darwin cheating Wallace out of the discovery, even incorporating illegitimately some of Wallace's work into his own. These claims have been found by nearly all historians of the field to be completely incorrect and unfounded.

Resolution

It is always possible that some unethical behavior may have occurred in cases such as this one. The evidence argues strongly against these claims, however. Wallace was given due credit at the time. The primary reason Darwin gets nearly all the credit today is likely the fact that he continued to contribute to the field while Wallace embarked on other pursuits. Furthermore, Wallace was never once reported to have taken issue with the outcome of the correspondence with Darwin. Finally, if Darwin wanted to, he could have just held Wallace's letter and never passed it on to Hooker and Lyell. The fact that he did not do this may stand as the most compelling piece of evidence in his favor. Furthermore, since 1908, the Darwin-Wallace medal has been awarded by the Linnean Society of London. Initially this award was issued every fifty years, starting in 1908. More recently (2010), the award has been issued annually. This suggests that, at least since the award's inception, fifty years after the work of Wallace and Darwin was reported, the scientific community has recognized the contributions of both men. That popular culture has all but forgotten Wallace is not Darwin's fault.

Questions to ponder

1. Was Wallace wise to contact Darwin and divulge what he did?
2. Did Darwin behave unethically?
3. Did Hooker and Lyell behave unethically?
4. How can we better recognize Wallace's contributions moving forward? Should we?

Sources

http//www.usatoday.com/tech/science/2009–02–09-darwin-evolution_N.htm (last accessed August 26, 2011).
http://en.wikipedia.org/wiki/Alfred_Russel_Wallace (last accessed July 5, 2011).
http://www.guardian.co.uk/science/2008/jun/22/darwinbicentenary.evolution (last accessed August 11, 2011).

Rangaswamy Srinivasan–VISX patent dispute

Summary

In 1983, Rangaswamy Srinivasan, a now-retired IBM research scientist, collaborated with Stephen L. Trokel, an ophthalmologist at Columbia University, to develop the technique that went on to become corrective eye surgery using lasers. In 1992, after their collaboration had ended for all intents and purposes, Trokel filed a patent, listing himself as the sole inventor. Srinivasan has not reportedly seen any of the financial benefits from his work. This is one of the most serious breaches of collaborative ethics in recent history.

The story

The ill-fated collaboration began in 1983 when Trokel convinced Srinivasan to help him (Trokel) develop a method for vision corrective surgery using an excimer laser. Not only did Srinivasan not know about a patent filed by Trokel in 1992, there were other issues too, from the earlier days of their collaboration, that would have caused a more cynical collaborator to walk away—something Srinivasan did not do. For example, a paper the pair co-authored was eventually found by Srinivasan to contain several errors. This paper, Srinivasan claims, was corrected by Trokel based upon suggestions from the *American Journal of Ophthalmology* editors. Among the errors Srinivasan found were a misuse of at least one reference and a referral to papers by Srinivasan that were already in print as "in press" or "unpublished." This last point is important since the timing of a report can be used to establish inventorship. In fact, he claims that not a single article that pointed in the direction of the phenomenon Srinivasan and colleagues at IBM had described was found in the paper. The phenomenon was critical to the success of the laser eye surgery method. In 2000, an International Trade Commission (ITC) ruling regarding the patent in question declared Trokel's patent invalid, claiming Srinivasan should have been named co-author. The ITC ruling goes on to state that Trokel knowingly and deliberately acted with deceit when submitting the patent. Other lawsuits have also been brought against Trokel and VISX, a company Trokel is a stakeholder in, regarding patents. Stockholders have sued the company, claiming that they had been misled by the fraudulent patents. Srinivasan has served in an expert witness capacity in several lawsuits brought against VISX and Trokel, and for his part only hopes that someday he receives due credit for his contribution to this work.

Questions to ponder

1. Should Srinivasan have been more vigilant in the corrections Trokel made to the manuscript they authored together?
2. Did Trokel really do anything wrong or was it Srinivasan's fault for not pursuing things, especially the patent, more aggressively?
3. Are the shareholders of VISX owed anything? After all, the company has made millions of dollars anyway.

Sources

W. G. Schulz, *Chemical and Engineering News*, May 28, 2001, 35–37.

Schwartz and Mirkin

Summary

This is the case of Peter Schwartz, former postdoctoral research associate in Chad Mirkin's lab, and Chad Mirkin. This case covers several issues, including: Who has the right to publish? What are the ethical issues of publishing without the consent of all co-authors and the failure to acknowledge the work of all co-workers? It is also highly important because it brings up other issues that do not quite fall into the category of any particular ethical violation. In particular, it involves the nature of the professional relationship between a research mentor/lab director and the junior scientists, and how their (or one individual's) professional interests may come into conflict with those of the university. There is a third individual involved with this story, Lydia Villa-Komaroff, the vice president for research at Mirkin's institution, whose comments are included below because they carry weight and provide the university's response to this issue.

How did it start?

Schwartz, after leaving Mirkin's lab, attempted to publish research in the journal *Langmuir* that he had performed while in Mirkin's lab. He tried to do so without acknowledging and without the consent of co-workers in Mirkin's lab and even and especially of Mirkin himself.

Mirkin says

Upon learning that Schwartz submitted the article to *Langmuir* and that the article had been accepted, Mirkin wrote a letter to *Langmuir* with several objections. These objections included that Schwartz:

1. Came into the lab with minimal expertise in the area of research.
2. Contributed to the group on one part of the project, then after some disagreements decided to leave the group and submit the entire research publication on his own.
3. Decided to submit the manuscript himself (without the consultation and without acknowledging co-authors)
4. Did not allow others in the lab to correct or verify the accuracy of the work.

The letter also claimed that:

1. There is some question regarding the interpretation of the work.
2. The work has never been reproduced by Schwartz or the Mirkin lab.

Schwartz says

Schwartz responded to several points in the following ways:

1. Schwartz received his PhD in 1998 from Princeton University, conducting research on the formation process of alkanethiol SAMs on Au (III)—the very system behind his research with the Mirkin lab that he was trying to publish and patent. He also worked with atomic force microscopy and scanning tunneling microscopy before joining the Mirkin group.
2–3. According to ACS author guidelines, he was the only significant contributor; however, he offered co-authorship to two members of

the Mirkin lab, an offer that was denied, with no one subsequently claiming authorship. He also claims that he offered co-authorship to Mirkin and that Mirkin responded by forbidding Schwartz from contacting him, insisting that all communication go through the vice president of research, who subsequently failed to respond to Schwartz's e-mails and calls.

4. The research for the manuscript took place after Schwartz left the lab.
5. Finally, Schwartz addresses the reproducibility by claiming that he has reproduced the results no fewer than fifteen times.

Mirkin responds

In response to the points made by Schwartz, Mirkin claims the following:

1. Schwartz was one person in a group of twenty-six on a project well underway prior to his joining the lab.
2. Schwartz had no prior experience with nanoparticles and DNA and his understanding of scanning probe microscopy was rudimentary compared to the group's.
3. It was made clear to Schwartz before his departure that no co-workers thought the work was publishable yet and that when it was, he would be offered co-authorship.
4. That Schwartz presented group ideas, developed over several years, as his own.

Villa-Komaroff's role

For her role in the matter, Villa-Komaroff states that she stopped correspondence with Schwartz when he informed her that he was retaining legal counsel. She consequently informed him that all correspondence must thereafter be through his attorney and the university's attorney. She has gone on to explain that the intellectual property (IP) actually belongs to Northwestern and that she had not released it, disallowing Schwartz from filing for a patent on the work.

Resolution

There has been a resolution, for now, regarding this case. The work in question was accepted for publication by the journal with an addendum stating that the scientist who supervised the work takes issue with some of its content, its ownership, and aspects of how the work was carried out. You can find the addendum at the end of the journal article in question. Schwartz believes that his results cast doubt on the performance of a competing technique developed by the Mirkin group.

Questions to ponder

1. Did Schwartz behave unethically?
2. Did Mirkin and his group behave unethically?
3. Did the journal behave unethically?
4. Are the university's Intellectual Property (IP) rules unethical?

Sources

Steve Ritter, *Chemical and Engineering News*, 2001, June 18, 40.

Chemical and Engineering News, 2001, July 30, 8–11; letters to the editor by Schwartz, Mirkin, and Villa-Komaroff.

David Adom, *Nature*, 2001, 412, 669.

Peter Schwartz, *V. Langmuir*, 2001, 17, 5971–5977.

Corey and Woodward

Summary

In 1965, Robert Burns Woodward was awarded the Nobel Prize for his outstanding achievements in the art of organic synthesis. In 1981, Roald Hoffmann, along with Kenichi Fukui, were awarded the Nobel Prize for their theories, developed independently, concerning the course of chemical reactions. In 1990, Elias James Corey was awarded the Nobel Prize for his development of the theory of methodology of organic synthesis. Although the pedigree presented here has no real bearing on the case presented below, it does provide perspective of the sheer magnitude and awesomeness of the characters involved. Corey claims that Woodward stole the idea that led to the Woodward-Hoffmann rules (and Hoffmann's Nobel) from him (Corey) during a conversation in Woodward's office. Because Woodward died in 1979, there is no rigorous way to refute or support Corey's claim, though anecdotal evidence does both. Corey repeatedly (both privately and publicly) called upon Hoffmann to set the record straight. In 1961, three years before this putative conversation between Corey and Woodward took place, a Danish chemist at Leiden University, L. J. Oosterhoff, published an article in the journal *Tetrahedron* first suggesting the very principles that later became the Woodward-Hoffmann rules. Woodward and Hoffmann correctly cite Oosterhoff's work, which applied the theory that they later generalized to a very specific system.

Corey says

In Corey's defense, his story has been unwavering for years. Although this by no means constitutes proof of truth, it certainly lends credence to his claim. He attested, during his Priestley Medal acceptance address in 2004, that he "suggested to my colleague, R. B. Woodward, a simple explanation … conversations that provided the further development of the ideas into what became known as the Woodward-Hoffmann rules." Corey communicated with Hoffmann often through the years, writing him letters throughout the 1980s, imploring him to set the historical record straight. These letters are available at the Cornell University library. One of the keys to Corey's argument is that Woodward (according to Corey) originally opposed the proposal that Corey made. The very next day, Corey claims, Woodward used Corey's explanation in a conversation with Professor Douglas Applequist, who was visiting. Woodward, according to Corey, presented the idea as his own, not crediting Corey, despite the latter's presence in the room (the conversation allegedly took place in Corey's office). Applequist, who knew Corey's interests in the area, later went so far as to express a level of surprise that Corey was not one of the co-authors on the original paper by Woodward and Hoffmann.

Hoffmann says

First and foremost, Hoffmann found Corey's attack on Woodward to be unfair at this point due to the fact that the latter has been dead for decades. Although Hoffmann claims he does not recall a conversation with Corey during the time in which Corey told him (Hoffmann) of the large role he played in Woodward's development of the idea, Hoffmann does admit that he asked Woodward if Corey should be included on the original paper proposing the idea. To this question, Woodward responded with a one-word answer: "No." The respect, admiration, and unwritten code of junior and senior colleagues compelled Hoffmann to not press the issue further.

Unlike Corey's account of the issues, Hoffmann's memory has not been as consistent, a fact he freely admits. This, however, does not mean he has been untruthful, nor does it mean he has been in error. Hoffmann has stated that, while he does not believe Corey's claims, if hard evidence could be presented, he would apologize and give credit to Corey for his seminal idea, setting the record straight in Corey's eyes. One final point that Hoffmann and others made in defense of Woodward is that it was not unusual for Woodward to ask questions of colleagues to which he already knew the answer.

L. J. Oosterhoff

Oosterhoff had no direct role in this controversy. His work, published in 1961 and cited by Woodward and Hoffmann in their paper, was the true first report of this theory. That they referenced it in their paper is proof-positive that they were aware of this prior work. To claim that Corey was unaware of this work by Oosterhoff would likely be a mistake; in all likelihood, he was indeed aware of it. Likewise, to claim that both Woodward and Corey were ignorant of this paper and that after the discussion with Corey, Woodward had performed a literature search and found Oosterhoff's paper is ludicrous. The year 1964 was a time of typewriters and telephones, not computers and the Internet. An overnight work session that would turn up such a paper and result in its thorough understanding is likely impossible, even for the best of the best, which Woodward arguably was. Furthermore, the likelihood that it was Corey who, with his suggestion, opened Woodward's eyes to this work is also low. If Corey's claim was instead that he gave Woodward the Oosterhoff paper as proof of his argument, the proof Hoffmann asks for may already have been there. Since this argument is not apparently Corey's claim, it makes it unlikely that Oosterhoff's paper was involved in the controversy.

Resolution

This is truly an immovable object vs. an unstoppable force. Corey is not going to convince Hoffmann his version of the events is true. The one man who can categorically refute or affirm Corey's claim has been dead for more than thirty years. In all likelihood, this controversy will not be resolved any further.

Questions to ponder

1. Should Corey have ever said anything?
2. What incentive would Corey have for waiting so long to voice these claims publically so late?
3. What, if anything, should Hoffmann do?
4. Should Hoffmann have pressed Woodward further?

Sources

http://www.boston.com/news/globe/health_science/articles/2005/03/01/ whose_idea_was_it/ (last accessed, August 26, 2011).
Chemical and Engineering News, letter to the editor, April 28, 2003.
Chemical and Engineering News, letter to the editor, February 8, 2005.
Chemical and Engineering News, letter to the editor, November 29, 2004.

Córdova, Scripps Research Institute, and Stockholm University

Summary

This is the case of Armando Córdova, at the start of this story a professor at Stockholm University. Córdova was found to be guilty of scientific misconduct on two out of four charges brought against him, and investigators also commented that many other allegations and rumors of scientific misconduct could not be fully substantiated. Here, the first two charges, those he was found guilty of, are discussed; the failure to cite previous work and attempting to pass another's work off as one's own. These are probably the two most prevalent forms of scientific misconduct. Another form of scientific misconduct that we will encounter in this story is the failure to publish without the consent of all of the researchers. Many different individuals will be mentioned in this story in addition to Córdova: Donna G. Blackmond, whose tireless pursuit to protect her work from theft brought this story to light; Carlos F. Barbas III, a former mentor of Córdova at the Scripps Research Institute; Stefan Nordland, dean of the Faculty of Science at Stockholm University; Olov Sterner and Torbjörn Frejd, independent investigators from Lund University; and Benjamin List of the Max Planck Institute, also a victim of Córdova's theft. One of the key points is that even after he was "busted," Córdova continued, rather than ended, his crooked tactics.

What happened?

The first documented incident involving Córdova occurred in 2001 when, while a senior postdoc in the Barbas lab, he submitted a research paper to the *Journal of the American Chemical Society* (where it was rejected) and then to *Tetrahedron Letters* (where it was accepted) without the consent of Barbas. Barbas was able to have the paper retracted after discovering what happened—by which time Córdova had been fired from the Barbas lab for reasons Barbas has not publically divulged. His incentive to have the paper retracted should be clear: the work was done under his supervision and then published without his consent. In early 2003, Barbas expressed concerns to Stockholm University that Córdova, where Córdova was employed, that Córdova might once again be attempting to submit the work for publication. Barbas's concerns were indeed well founded as Córdova submitted the paper the following month to *Synlett*, where it was accepted with Córdova as the sole author. The Sterner-Frejd investigation ruled that this was "a clear case of unethical behavior." In 2003, *Synlett* published an addendum giving the history of the manuscript in question, including the comments from the Sterner-Frejd investigation that found

Córdova guilty of unethical behavior with respect to this manuscript. The editors in the addendum state, "It is much to the regret of the Editors and the Publisher of *Synlett* that the paper has been published in *Synlett*."

Another incident in which the Sterner-Frejd investigation found Córdova guilty of scientific misconduct was the case involving Donna Blackmond of Imperial College in London. In late 2005, Blackmond delivered the Holger Erdtman Lecture at KTH, the Royal Institute of Technology in Stockholm, a lecture that Córdova attended. The work that Blackmond presented had been recently submitted to *Nature* for publication. A later investigation determined that during the weeks after the lecture, Córdova's group studied a similar system and attempted to publish the results in *Chemistry—A European Journal*, a venue known for rapid publication. This paper was accepted for publication and eventually read by Blackmond. In reading Córdova's paper, Blackmond recognized some of the concepts that were developed by her lab and that her work had not been cited. Blackmond went on to point out multiple weaknesses in Córdova's paper and requested that he be required to publish a corrigendum. Córdova counterclaimed that Blackmond had actually gained insights from one of his previous papers that she had reviewed and that she failed to cite his work. Córdova's paper that he refers to as not being cited by Blackmond was rejected for publication, making it impossible to have been cited, and he later withdrew his allegations. The Sterner-Frejd investigation found evidence to support Blackmond's claims when they examined Córdova's lab notebooks. Finally, while Blackmond and Córdova settled the issue of a corrigendum to the *Chemistry—A European Journal* publication, Córdova submitted some of the same data and similar claims as in that same *Chemistry* paper to another journal, *Tetrahedron Letters*. This paper was accepted despite not only its similarity to already published and under dispute work but also despite not citing Blackmond's work properly. In this new paper, Blackmond's work was cited, to be fair to Córdova, but it was buried deep in the paper and failed to give Blackmond appropriate credit for being the first to report the observed behavior. As if these transgressions were not enough, Benjamin List of the Max Planck Institute for Coal Research in Mülheim, Germany, claims that, after speaking at a conference in Italy that Córdova attended, similar work to List's, published by Córdova, appeared in *Tetrahedron Letters*.

The penalty that Córdova received was that he had to attend an ethics course and that all of his papers had to be presented to his dean before publication. Nordland (Córdova's dean) stated that he feels that this penalty is sufficient, while also commenting that he "is very unhappy with and do[es] not support the behavior of Córdova" but going on to claim that "this is not the worst ethical behavior in science," in partial response to potentially firing Córdova for his scientific misconduct. Córdova, in his own defense, has steadfastly claimed that he didn't know this sort of behavior was foul and that he was just following the mentoring of former advisors, a claim that Barbas has responded harshly to.

Resolution

To date, there have been no further officially publicized resolutions to this ordeal. Browsing the totallysynthetic.com blog, there have been unverified rumors that Córdova has been blacklisted by the ACS journals. Searching the ACS publication website for Córdova gave no hits after 2006, the year that many of these incidents occurred, lending credence to these rumors. With that being said, Córdova certainly does not seem to be doing too badly for himself, career-wise. He is currently a professor in organic chemistry and a researcher at Mid Sweden University and Stockholm University while also serving as the head of the Chemistry Department at Mid Sweden University. Furthermore, in 2009, Córdova was selected for a professor chair in organic chemistry.

Questions to ponder

1. Was what Córdova did truly wrong?
2. Were the penalties levied on Córdova appropriate?
3. Should an entire family of journals blacklist individual authors the way the ACS appears to have done to Córdova?
4. Was Barbas wrong in trying to prevent Córdova's publication?
5. Did Blackmond overreact?

Sources

Chemical and Engineering News, 2007, March 12, 35–38.
http://Armandocardova.com (last accessed June 22, 2011).
http://totallysynthetic.com/blog/?p=322, found when searching Google for Córdova and Blackmond (last accessed June 22, 2011).
Synlett, 2003, 2146.

La Clair and hexacyclinol

Summary

The case of James La Clair (then of the Xenobe Research Institute) is a curious one. There has been no formal charge of misconduct and it is unclear there ever will be, though there have been half-hearted accusations. At the heart of the story is the structure of hexacyclinol, a complex natural product La Clair claims to have synthesized.

What happened?

La Clair reported in 2006 the total synthesis of hexacyclinol with the original structure solved by Udo Gräfe,[1] the original reporter of the structure of this antiproliferative compound. Later, in 2006, Scott D. Rychovsky of the University of California at Irvine proposed a different (and later verified structure) based on high-level ^{13}C nuclear magnetic resonance (NMR) calculations. Then later that same year, the lab of John Porco at Boston University synthesized the revised Rychovsky structure, proving it was indeed the correct structure; this synthetic sample matched the authentic isolated sample in every way. In La Clair's report, the ^{1}H NMR was also found to be identical but the ^{13}C NMR contained some inconsistencies, which La Clair attributed to the different solvent used to collect his spectrum.

What make this story intriguing were the circumstances surrounding the La Clair synthesis. For example, he did not include the five technicians who assisted him as co-authors on the paper reporting the synthesis, claiming that he "didn't mean to offend anyone" and "if the editors or reviewers had told me that was unethical, I would not have done it." There are also steps in La Clair's synthesis that are very questionable, the details of which are beyond the scope of this book. In any case, it causes some to question whether he had actually synthesized the product. To be fair, at least one of these steps seems so intricate (adding a reagent once an hour for five hours) that one can only conclude it either really did happen or that La Clair has the best imagination on Earth. A more careful look also reveals that much of the work was done while La Clair was at Bionic Bros. GmbH in Berlin, and acknowledging technicians and not including them as authors was not uncommon in Germany. However, other oddities surround this work as well. For example, La Clair comments that the ^{1}H NMR contained a solvent peak added by the operator (whatever that means) against his (La Clair's) wishes. As if that were not strange enough, the solvent peak in question was not added at the right position. It is unclear why La Clair chose to continue to use this NMR service (his lab did not collect the data itself and instead used a paid service).[2]

Resolution

It appears that this case will proceed no further. It is unlikely that formal charges of fraud will ever be pressed. It is possible (though unlikely)

---

[1] Remember from our discussion of proving previous results wrong, natural products will occasionally have their structures misassigned.

[2] This is not unethical.

that La Clair's structure and the correct structure would have the same ¹H NMR spectrum. It is also possible (though perhaps even less likely) that an error occurred in La Clair's synthesis that actually caused him to unknowingly produce the correct structure. It is also possible (with unknown likelihood) that La Clair or the NMR service he used fabricated results. I am by no means intending to accuse anyone of misconduct, merely acknowledging all possibilities. Finally (and probably most likely), some other kind of error was made and, as usual, science ferreted it out and fixed it.

Questions to ponder

1. Was there any scientific misconduct?
2. Was there any bad science?
3. What could be done to check La Clair's original synthesis?
4. Did the NMR operator behave unethically?

Sources

B. Halford, *Chemical and Engineering News,* July 31, 2006, p. 11, and references cited therein.

B. Halford, http://cenonline.blogs.com/sanfrancisco_2006/2006/09/hexacyclinol_ sh.html (last accessed August 26, 2011).

http://www.thechemblog.com/?p=210 (last accessed August 11, 2011).

http://pipeline.corante.com/archives/2006/06/05/hexacyclinol_or_not.php (last accessed April 22, 2007).

http://www.thechemblog.com/?p=108 (last accessed April 22, 2007).

Woodward and quinine

Summary

Quinine is a potent antimalarial drug (and for a while was the only recognized one). Its widespread need was perhaps greatest during World War II when access to the cinchona trees was blocked by the Japanese takeover of Java. Robert Burns Woodward and William von Eggers Doering answered the urgent call for a synthesis, though their work was never actually used to prepare this important drug. For reasons unrelated to the story that follows, Woodward and Doering never actually prepared quinine with this process.

What happened?

Woodward and Doering titled their paper "The Total Synthesis of Quinine." In fact, their synthesis was *not* a total synthesis by today's standards; instead, it was what we call today a formal synthesis. It should be noted that the term *formal synthesis* was not widely used in 1944, the time of their publication, though it is certainly used today. These semantics have no relation to whether Woodward and Doering behaved unethically; they used terminology consistent with their era. It would therefore be a thorough waste of time to get hung up on their choice of words. What is the difference though? A total synthesis (in today's terms) is quite self-explanatory: It is the preparation of a target (usually a natural product) from commercially available starting materials. It is a *total* synthesis, something made from scratch. A formal synthesis (again, today's terms) is a synthesis that arrives at an intermediate state that someone else has shown can be taken to the final product. The latter is what Woodward and Doering did. In their paper presenting their work, Woodward and Doering properly cited the work of Rabe, whose work in 1918 showed that the same intermediate they made could be taken to quinine. Several synthetic organic chemists since that time (among them, Gilbert Stork, who himself contributed the first stereoselective total synthesis of quinine in 2001) questioned whether or not Rabe's synthesis would have worked at all. Although Hoffman-LaRoche briefly explored Rabe's approach in the 1960s, they found it to be in need of major alteration, which made it impractical for them, and they published their own synthesis of quinine in the 1970s.[3] It was not until 2008 that researchers finally put the Woodward-Doering-Rabe method to a full test. Robert Williams and Aaron Smith at Colorado State in Fort Collins demonstrated Rabe's synthesis worked reasonably well until the last step. They ultimately found that if one of the reagents in this step was exposed to air (conditions highly possible if not likely in Rabe's more primitive 1918 German lab), the troubling reaction worked quite well.

Resolution

Even Stork, who doubted Rabe's approach and argued strongly to have the record set straight with Woodward and Doering's synthesis (all the while also admiring what the pair did), praised the work of Williams and Smith. Again, here, there may not be more to discuss in the way of misconduct. No misconduct was ever accused or even hinted at. Instead, what was subtly challenged is the norms of science. Is it "right" to take someone else's word for experiments they have performed? If we can't, the whole system falls apart because we are then forced to constantly re-invent the wheel. But if we always do and never check, the self-correcting nature of

---

[3] This should not be interpreted as an indictment of the validity of Rabe's method. When pharmaceutical companies scale up production, they often have to greatly revise the syntheses.

science never operates. Often, any problems would have been found out if the Woodward-Doering intermediate and Rabe's synthesis had been used commercially, but by the time of publication of Woodward and Doering's work (1944), the war was nearly over, easing the need to explore this route.

Questions to ponder

1. Did Woodward and Doering act unethically by not completing the synthesis?
2. Did Woodward and Doering commit bad science by not completing the synthesis?
3. What were Stork's motives for being so vocal?
4. Should Hoffman-LaRoche have revealed the problems they encountered?

Sources

A. M. Rouhi, *Chemical and Engineering News*, May 7, 2001, 54–56.
Chemical and Engineering News, letter to the editor, June 18, 2001.
Chemical and Engineering News, letter to the editor, August 31, 2001.
B. Halford, *Chemical and Engineering News*, February 26, 2007, 47–50.
B. Halford, *Chemical and Engineering News*, February 4, 2008, 8.

DNA

Summary

The story that surrounds the elucidation of the structure of DNA is one of the most intriguing in modern science. The importance of the work that revealed the chemical structures of the molecule of life is impossible to overstate. It is, in fact, difficult to think of ten discoveries that supersede its importance (at least when you consider how many important breakthroughs were possible because of it). What many people do not know, however, is that an enormous level of controversy surrounds this monumental discovery. Few controversies bring about as much of a tempest as this one does, in the right company. The cause for this? Rosalind Franklin and the way she and her work was (and to some, still is) perceived to have been treated. To be fair, there is at least some evidence she may have gotten the short end of the stick. Unfortunately, the world will never know if Franklin would have shared the Nobel Prize with James Watson and Francis Crick (presumably at the expense of Maurice Wilkins); she died in 1958, before this trio was chosen for the award (1962). The award cannot be given to deceased individuals nor can it be awarded to more than three people. The argument can certainly be made that Wilkins contributed significantly less than Franklin did and if she had lived, she would have gotten the award as a result. Awarding it to Wilkins perhaps can actually be perceived as a nod to Franklin. Alternatively, Watson and Crick might have received one and Wilkins and Franklin another. We will simply never know.

What happened?

In 1951, James Watson arrived in Cambridge, in the Cavendish Laboratory, and was immediately teamed up with Francis Crick by Sir Lawrence Bragg. Thus began a friendship and work partnership that would quite literally change the world. At this early point, it is worth pointing out, at least with regard to the elucidation of the structure of DNA, that neither Watson nor Crick collected a single piece of experimental data themselves for this project. This undoubtedly is a partial source for the bile frequently spewed at them. However, this should take nothing away from the feat they accomplished. They put the pieces together (figuratively and literally) in ways nobody had previously done. Through attending seminars and meetings they legitimately (at least at first) acquired data that led them eventually to the right answer. Their primary source of DNA x-ray data was Rosalind Franklin and Maurice Wilkins, who were co-workers of one another at King's College in London.

At this point, it should be mentioned that Wilkins and Franklin had a relationship that was several steps beyond strained. The blame for this, at least in part, was John Randall, head of the Biophysics Unit (Franklin and Wilkins's unit) at King's College. Randall, while interviewing Franklin, explained to her that she would be working independently, while also intimating to Wilkins that Franklin would be working under him (Wilkins). It is no small wonder, with this in mind, why the two were more in opposition to one another than true colleagues should ever be. As for Randall's motive, one can only speculate now. It has been supposed that with Franklin and Wilkins rivals, Randall would be more able to sweep in and share credit if either made a major discovery. Although some have wondered if the two would have ever truly coexisted working together even without Randall's stoking the fire, it is impossible not to speculate. If they had been able to work cooperatively, perhaps they would have been able to beat Watson and Crick to the goal.

In 1952, Watson attended a seminar at King's by Franklin, an update on the last six months of her work. However, he was sans notebook, as he often relied on a very keen memory. Much to especially Crick's dismay, Watson flubbed several details, including the number of water molecules (he was off by an entire order of magnitude). What followed was an unmitigated disaster for Watson and Crick. They built a heinously incorrect model based upon Watson's recollection of Franklin's seminar.

Making matters worse was their invitation to Wilkins and Franklin to come view their model for approval. Franklin immediately began pointing out flaws in the proposed structure, with the number of waters it contained being just one of her complaints. To say that she was less than impressed with the degree to which her results had been misinterpreted would border on a comical understatement. Not surprisingly, word of this episode eventually reached Bragg, causing him significant embarrassment. His impatience with Crick was already thin for unrelated reasons, and this perceived besmirching caused him to forbid the pair from further work on DNA. At least part of his argument was that he considered it ungentlemanly and against the norms of British science to work on something another Brit was studying. Such toe-stepping was to be avoided, in his opinion.[4] This could have been the end of Watson and Crick's contribution to the story of DNA, and probably would have been if Lady Luck (or perhaps it was fate) had not intervened later in 1952, shortly after the episode with Franklin and Wilkins. Peter Pauling, son of Linus Pauling, one of the greatest chemists to ever live, joined Cavendish. The importance of this was that Watson and Crick had long feared the elder Pauling would eventually focus his considerable intellect on elucidating the structure of DNA, for he did related work. Later that year, a letter from father to son that the younger Pauling shared with his co-workers confirmed these fears. The great Linus Pauling was hot after the structure of the molecule of life, and Watson and Crick effectively began an imaginary countdown to the day Pauling would announce that he had done it. During this time, Franklin's x-ray pictures of crystallized DNA obtained from Wilkins were being collected with ever-increasing quality. Her photographs were even higher quality than anything Pauling had. One picture in particular clearly showed the helical nature of DNA. Despite this, Franklin reported in an official in-house lab announcement that the helical structure was incorrect. Wilkins, however, was a supporter of the helical structure.

Early in 1953, the Pauling paper announcing the structure of DNA was released and Watson and Crick were immediately relieved. His structure actually bore a remarkable resemblance to the failed structure Watson and Crick themselves proposed in 1952 just before Pauling's son joined Cavendish. Pauling got it wrong. Convinced it was only a matter of time before he discovered and corrected his error, Watson and Crick resumed work (covertly at first) on DNA. The truth is, they never stopped thinking about it, just stopped openly working on it. But, with this gaffe by Pauling, they decided they had to act quickly and deliberately. Watson went to King's College. He was to meet with Wilkins but arrived early and decided to stop and see Franklin to discuss Pauling's error. When Watson began harping on his and Crick's preference for the helical, and

---

[4] Remember this point later on!

maybe double helical, structure, Franklin became enraged—frighteningly so, if you believe Watson's account. This does have some import here. Previously, Franklin had declared there to be no helix, despite evidence to the contrary. Her reaction here suggests she really might have believed her interpretation to be true. If this is the case, she almost certainly allowed her friction with Wilkins to cloud her interpretation, since Wilkins was also a supporter of the helical structure. If Wilkins and Franklin had been more teammates than adversaries, perhaps it would have been they who shared the prize.[5] In any event, Watson was "saved" from Franklin's rage by the appearance of Wilkins and the two retired to Wilkins's office. Here, Wilkins shared with Watson a copy of one of Franklin's x-ray photos. This photo contained remarkable clarity and, for Watson, absolutely confirmed the helix. Unbeknown to Wilkins, Watson and Crick would soon obtain that same data (including some of Wilkins's own data) from other means as well. One of Watson and Crick's colleagues, Max Perutz, was part of a committee appointed by the Medical Research Council, to which Wilkins and Franklin's results were sent, and Perutz shared these with Watson and Crick. Eventually, Watson and Crick were no longer able to hide their resumed pursuit from Bragg. Fortunately for them, Bragg and the elder Pauling were rivals. With Pauling now so obviously in very hot pursuit, the gentlemanly behavior Bragg was so beholden to earlier evaporated. Watson and Crick could now work on this monumental project with his full blessing. Moreover, they were permitted to employ the machine shop to build their model parts.

The next great epiphany came to Watson from the Chargaff rules that there are equal amounts of the bases adenine and thymine, and equal amounts of the bases cytosine and guanine, in DNA. The clarification from Jeremy Donohue about the structures of guanine and thymine in a discussion with Crick was the final piece of data they needed. All of this data combined made the formal structure assignment academic for Watson and Crick. They assembled their model and the rest is history. Shortly thereafter, Wilkins and Franklin visited and approved the structure. The foursome agreed to publish all their research in series in the same issue of the journal *Nature*. A little while after that, Linus Pauling stopped in and agreed that Watson and Crick had the correct structure.

Resolution

Only Watson remains alive from the main players in this story. He has claimed that if Franklin had lived, she would have shared the Nobel.

---

[5] The intensity of the friction between these two is likely all three of their fault (Randall, Wilkins, and Franklin). The comments herein are not meant to vilify one person.

Whether this would have been at Wilkins's expense (which would have probably been appropriate since, although Wilkins eventually collected his own data, Franklin's may have more directly led to the elucidation) or on her own with another Nobel Prize is unclear. Watson's opinion appears to be that it would have been at Wilkins's expense. Many of the people involved with this story acted like children at times. Watson claims that after the affair was over, Crick enjoyed a friendly relationship with Franklin but at this point, he can only offer a secondhand account that cannot be corroborated.

Questions to ponder

1. Did Bragg behave unethically in forbidding Watson and Crick from working on DNA?
2. Did Bragg behave unethically in changing his mind when allowing Watson and Crick to resume work?
3. Did Wilkins behave unethically in sharing Franklin's data with Watson?
4. Did Randall behave unethically in setting Franklin and Wilkins against one another?
5. Did the younger Pauling behave unethically in sharing the letter from his father?
6. Did Watson and Crick behave unethically in not collecting their own data?
7. Did Peretz behave unethically sharing the x-ray data?

Sources

M. White, *Rivals: Conflict as the Fuel in Science*, Vintage Books, 2003, 231–273.
J. D. Watson, *The Double Helix: A Personal Account of the Discovery of the Structure of DNA*, Touchstone Books, 2001.

David Baltimore and Teresa Imanishi-Kari

Summary

This case is peculiar in the sense that the conclusions that were made from the fabricated data were eventually proven true by other studies. This has led some (including one of the authors of the study) to claim that it does not matter that data was fabricated. In this case, a postdoctoral researcher (O'Toole) brought to light results fabricated by her mentor (Imanishi-Kari) who was involved in collaboration with the principal investigator (PI) and Nobel Laureate (Baltimore). This case is filled with issues of fabricated data and other forms of deception, sloppy science, trouble for the whistle-blower, and even a full-blown congressional investigation.

O'Toole's side

Margot O'Toole began working with Professor Imanishi-Kari at MIT in 1985. During her tenure there, part of her task was to replicate results Imanishi-Kari had previously observed. After many trials, O'Toole was unable to replicate some results when she worked on her own, causing Teresa Imanishi-Kari to have several allegedly belligerent and emotional blow-ups. O'Toole was eventually assigned the task of maintaining the mouse colony to finish off her term in the lab. At one point, her new task required her to check breeding records to verify the pedigree of a particular mouse. What she found stunned her. First, the records appeared devoid of any organizational scheme. Second, and this is what ignited the controversy, O'Toole stumbled across data that agreed with the result she (O'Toole) was obtaining; there was no indication that Imanishi-Kari's result had ever been observed. The actual results were, to O'Toole's figuring, misrepresented in a paper Imanishi-Kari published with David Baltimore in the journal *Cell* that presented the results O'Toole could not replicate, though she did obtain the "correct" result (the one in the *Cell* paper) once, when assisted by a junior colleague of Imanishi-Kari and the progenitor of the work reported in *Cell*. O'Toole further asserted that the data she found in the books showed that there were defects in the study that were not acknowledged in the published manuscript.

Just two days after the discovery, O'Toole brought her concerns to a trusted researcher at Tufts University, Brigitte Huber, who counseled O'Toole to take the concern to Henry Wortis, who was supporting Imanishi-Kari's impending move to Tufts. As a result, Huber and Wortis, along with Robert Woodland, met with Imanishi-Kari. The trio concluded after the meeting (which O'Toole was not party to) that nothing would be done to correct the paper. Understandably, O'Toole was not satisfied with this response and took her concern to Martin Flax, chairman of the Pathology Department at Tufts. Flax's response, considering Imanishi-Kari was in the process of moving to his department, was baffling: He informed O'Toole that the issue was MIT's problem. Huber and Wortis then met again with Imanishi-Kari, this time along with O'Toole. At this meeting, Imanishi-Kari produced two pages of data. During this time, O'Toole had also met with Gene Brown, dean of science at MIT, who told O'Toole to formally charge fraud or drop the matter entirely. Brown contacted Herman Eisen, a collaborator on a grant with Imanishi-Kari that the funding agency dropped Imanishi-Kari from, and requested he contact O'Toole. When they met, O'Toole presented the seventeen pages she found to Eisen, who, according to O'Toole, responded, "That's fraud."

Eisen has since claimed he does not remember this and accuses O'Toole of being incoherent during the meeting.

Another meeting, this one arranged by Eisen, between Baltimore, another researcher David Weaver, Imanishi-Kari, and O'Toole, took place a short time after. This new meeting produced no real changes—nothing was to be done.

Charles Maplethorpe

In the summer of 1986, the case took on a life of its own and out of the hands of O'Toole, at least partially. Charles Maplethorpe, a former student of Imanishi-Kari with a history of friction with her, brought the case out into the public eye. The result was a congressional hearing led by Joseph Dingell, a representative from Michigan who was also pursuing other cases of fraud in science and has also done so since that time. At the hearing, Maplethorpe testified that he overheard a conversation between Imanishi-Kari and another researcher, David Weaver. During this conversation, Imanishi-Kari expressed she had been having trouble with a particular reagent and was seeing the same result O'Toole would later observe. To Maplethorpe's knowledge, this issue was not resolved prior to the publication in *Cell*. Imanishi-Kari protested this account on the grounds that Maplethorpe was not a disinterested witness.

In response to the congressional hearing, Imanishi-Kari produced a looseleaf notebook she claimed contained original data later organized into that form. At this point, O'Toole finally formally charged fraud, concluding, along with others, that some pages were faked. Meanwhile, an NIH-appointed panel commenced its own investigation, appointed by then-director James Wyngaarden. The panel was chaired by Joseph Davie and although it found serious flaws in the paper, it did not classify anything as misconduct. O'Toole insisted this finding was in error. Wyngaarden reopened the investigation, appointing a new panel and establishing an Office of Scientific Integrity.[6] Regarding the previously mentioned notebook, O'Toole's contention that it was faked was corroborated by the forensic services division of the Secret Service. Their examination concluded that the notebook could not have been produced at the same time as the work.

Imanishi-Kari was eventually found guilty of two counts of fraud. The first was in the original paper in *Cell*, the other was in the production of the notebook. The final report, issued in 1994 by the Office of Scientific Integrity, detailed eighteen charges of scientific misconduct. The penalty was proposed to be a ten-year ban on applying for federal

---

[6] This was the precursor to the Office of Research Integrity, demonstrating the long-lasting impact of this case.

grants. This would have included not just NIH, but the National Science Foundation (NSF), the Petroleum Research Fund (PRF), the Department of Energy (DOE), and so forth. All federal monies would be unavailable to Imanishi-Kari for ten years—a career-killing punishment. She was also suspended from Tufts. She appealed to the Department Appeals board of the Department of Human and Health Services, and the panel in the Office of Research Integrity ruled that the original case did not prove its charges by a preponderance of evidence, and so Imanishi-Kari was again allowed to seek federal monies and her suspension at Tufts was lifted. The decision went on to criticize the Office of Research Integrity and O'Toole as well.

Teresa Imanishi-Kari

Imanishi-Kari was not new to controversy when the case erupted. One previous incident involved her claims that she holds a master's of science degree from the University of Kyoto. The *Boston Globe*'s Tokyo Bureau investigated and found this to be false. Two members of the Laboratory of Developmental Biology at the University of Kyoto apparently once wrote a letter stating, "from these evidences it may be evaluated that her two years activities are equivalent in quality to complete two years master course at our University." The head officer of sciences at the University of Kyoto, Yoshio Mitabe, wrote in 1995 during an Office of Research Integrity investigation that "we could not find any official evidence that she was enrolled or employed in any position or grade at Kyoto University," going on to refer to the letter that Imanishi-Kari produced as a "private-level certification without administrative will." This becomes significant in that the University of Helsinki, where she received her PhD, mandated that foreign students with no more than a bachelor's degree would still be considered to be undergraduates. To gain admission into its doctoral program, the student must have already possessed a master's degree. The controversy about the master's degree, or lack thereof, could also have landed Imanishi-Kari in hot water with federal agencies because it could be found to be a falsification of her qualifications and professional biography. It is also important to note that the National Cancer Institute recommended full funding for a joint grant—except for the work to be done by Imanishi-Kari's lab. This grant was to work on the very paper in dispute, discussed here. This blow came shortly before the paper was submitted to *Cell*, so in all likelihood, the grant would have funded the next steps. Imanishi-Kari was also reputed to be a poor record keeper, someone who did not see experiments through, and someone who sloppily did calculations in her head, sometimes making errors, off by an order of magnitude. The adage "where there's smoke, there's fire" could well apply here.

David Baltimore

This man, David Baltimore, is perhaps the reason this case took on the life that it did. Baltimore is a Nobel Laureate and thus a giant in his field. It is difficult to say that this case would have been afforded the same level of attention if it had not involved someone of his eminence. As the case progressed, Baltimore was appointed the president of Rockefeller University. By the time it mercifully ended, he was forced out of that position. It should be clear that at no point was Baltimore accused of fraud. Belligerence, at times, yes—but never fraud. Although the reality is that all researchers are responsible for the work being reported, a good liar can get nearly anything past his or her collaborators. This appears to be what Imanishi-Kari did to Baltimore, though it can certainly be argued that Baltimore's fame is why he was not also accused of fraud. One particularly troubling comment that Baltimore made, however, bears mention. Baltimore contends that since independent studies have more or less led to the same conclusions, it is not a big deal since the fraud clearly did not hold the science back. He also has allegedly wavered on his stance regarding the importance of the fabricated work's conclusions—on the one hand being reputed as implying that they are the whole point and then publically claiming they are minimal in importance.

The public perception

The perception of this case, in particular within the scientific community, was curious. At one point during the congressional hearings, scientists were called upon by Philip Sharp, an MIT scientist and friend of Baltimore, to write their congresspersons and submit op-ed letters to their local newspapers railing against the hearings, perceived by some to be nothing more than a witch hunt, an intrusion into science by lawmakers. This is an unfair characterization of the zeal with which Dingell pursued this case and others during that time. Dingell certainly appeared incorrigible. However, the federal government should investigate with rigor accusations of fraud in cases that involve federally funded research. Any congressional investigation must heavily rely on the opinion of research scientists. With this qualifier, the decision can be made regarding official penalties with an informed mind. This says nothing about, nor does it have any influence over, the unofficial penalties that ultimately are levied upon scientists who are perceived to commit fraud. Often, this stigmatization follows them forever and since the author of a grant or manuscript is rarely anonymous,[7] it inevitably affects the review process, unleashing the self-policing power of science.

---

[7] Such reviews are called double-blind. In the double-blind reviews, the reviewer is not told who the author of the grant or manuscript is.

Conclusions

Out of necessity, many of the finer details were left out here. The interested reader is encouraged to read the account by Horace Freelance Judson, who includes firsthand observations and conversations in his narration. Enough information is hopefully included here to stimulate a provocative discussion of what was right and what was wrong. When all that occurred is analyzed, one cannot help but conclude that Imanishi-Kari got away with her misconduct. Baltimore, if he is guilty of anything, is guilty of allowing her to do so. O'Toole, the steadfast defender of what was widely agreed as being the truth, displayed much of what is good and right about science. The Dingell investigation demonstrated both why Congress should get involved and, with what was later perceived to be a bungled investigation, why it should not.

Questions to ponder

1. Should Baltimore have felt more heat?
2. What could Baltimore have done differently?
3. Did O'Toole do anything wrong?
4. Did Maplethorpe do anything wrong?
5. Was Sharp's call to scientists appropriate or inappropriate?

Sources

H. F. Judson, *The Great Betrayal: Fraud in Science*, Harcourt, 2004, 191–243.

John Fenn–Yale patent dispute

Summary

John B. Fenn shared the Nobel Prize for chemistry in 2002 with Koichi Tanaka and Kurt Wurthrich for his work in the development of electrospray ionization for the analysis of large molecules. Fenn filed patents on his work, and in 2003 courts ruled in Yale's favor that the grant should be Yale's property.

The story

Fenn, in the late 1980s, developed the electrospray ionization mass spectrometry method, allowing the mass spectroscopy of large molecules to be measured reliably for the first time. In 1992, Fenn applied for and was awarded a patent on the method, with himself as the assignee, meaning he would be the principal earner of any monies gained. This is very atypical for scientists, especially at academic institutions. He then licensed the patent to a company he cofounded, Analytica, of Branford, Connecticut, who then sublicensed rights to instrument makers. This last part is a typical protocol. By this time, Fenn had relocated from Yale to Virginia Commonwealth University, having been forced into retirement by Yale.

When Yale discovered this patent, it claimed rights to the patent, asking Fenn to reassign the patents to the university, a request Fenn refused. Fenn subsequently sued Yale and Yale filed a counterclaim, claiming that Fenn "misrepresented the importance and commercial viability of the invention … [and] actively discouraged Yale from preparing and filing a patent application," going on to claim that Fenn inappropriately filed the patent himself, neglecting to notify Yale or the National Institutes of Health (NIH), which provided funding for the work. Fenn defended himself, claiming that Yale did not pursue a patent because it did not view it to have sufficient commercial interest, changing course when the university realized it was earning Fenn substantial income. He pointed out that earlier electrospray patents were assigned to Yale and have been valuable to the university.

If, as Yale claims, Fenn undersold the potential value of the patent to the university, this would certainly account for Yale's decision to not pursue a patent. In 2005, a court gave its final decision, ordering Fenn to pay nearly $1 million in combined misdirected royalties and legal fees and that the patent be transferred to Yale, convicting Fenn of civil theft. In late December 2010, Fenn, ninety-three years old at the time, passed away.

Questions to ponder

1. If Fenn did undersell the value of the patent to Yale, did he commit misconduct?
2. If Yale exaggerated Fenn's initial claims regarding the commercial value of the work, should they have been awarded the patents?
3. Is it at all unethical that the university receives a significant cut of patent royalties?

Sources

http://en.wikipedia.org/wiki/John_Fenn_(chemist) (last accessed September 22, 2011).

http://www.washingtonpost.com/wp-dyn/content/article/2010/12/11/AR2010121102387.html (last accessed September 22, 2011).

S. Broman, *Chemical & Engineering News*, February 21, 2005, 11.

VIOXX®

Summary

In 2004, pharmaceutical giant Merck voluntarily removed its painkiller VIOXX from the market. It was found during the course of a three-year study on some other activity of the drug (after the FDA approved it as a painkiller) that the drug placed participants at higher risk of cardiovascular events. If Merck's records are correct, and it is important to note that there is little reason to doubt they are, the side effects were not known when the FDA approved the drug in 1999. It is important to know that all the data Merck would have collected during clinical trials would have been shared with the FDA in order to gain approval, meaning Merck would have had to falsify the results sent to the FDA to get VIOXX accepted amid serious side effects. It is also in Merck's favor that the withdrawal was voluntary. They found a problem and took the appropriate steps to rectify it. As a result (at least in part) of the issues Merck uncovered with VIOXX, one of Merck's competitors, Pfizer, initiated a large study to investigate its own drug, Celebrex, which is a member of the same general family of drugs as VIOXX.

It is certainly not unheard of for drugs to be used and eventually approved for new treatments. Therefore, Merck explored this for VIOXX and its activities were not at all unethical; in fact, they are quite typical. A perfect example of this is aspirin. Most of us grew up using aspirin as a painkiller. Today, it is also approved for treatment during a heart attack and after suffering a stroke. For years, it was informally used for these additional purposes. Aspirin has been around a very long time, much longer than VIOXX, so its safety record was already well established, making it easier for aspirin's uses to be enlarged. Pharmaceutical companies often re-deploy their drugs for additional use. Sometimes, some of the uses are unofficial, occasionally staying that way. Others eventually gain multiple approvals.

This case is different from the others in this chapter. This case has been deliberately included to open up the discussion about drugs being used for multiple purposes, including what dangers may be associated with this. These uses are referred to as "off-label" uses when they are not approved uses. A list of drugs, their approved uses, and their off-label uses are provided in Table 8.1.

Question to ponder

1. Should it be allowed that drugs can be used for any purpose other than the one they went through clinical trials for?

Table 8.1   Off-Label Drug Use

Name of drug	Official use	Off-label use
Methotrexate	Choriocarcinoma	Unruptured ectopic pregnancy
Sertraline	Antidepressant	Premature ejaculation in men
Adderal and Ritalin	ADD in children	ADD in adults[a]
Gabapentin	Seizures and post neuralgia in adults	Bipolar disorder, essential tremor, hot flashes, migraine prophylaxis, neuropathic pain syndrome, phantom limb syndrome, and restless leg syndrome
Viagra[b]	Erectile dysfunction	Pulmonary hypertension

[a] Yes, it matters that they're different!
[b] An interesting note about Viagra: Viagra was initially intended to be a heart medication. During clinical trials a side effect was reported by a large number of men in the study. Pfizer quickly realized that this side effect, if it were a bona fide result, would make them much more money than the original use, and so they revised their application and likely patents to reflect this. Millions of dollars later, the rest is history.

Sources

V. Marx, *Chemical and Engineering News*, October 4, 2004, 8.
V. Marx, *Chemical and Engineering News*, October 25, 2004, 15.
R. Mullin, *Chemical and Engineering News*, November 15, 2004, 7.
Chemical and Engineering News, letter to the editor, January 3, 2005.

Index